Th

About the Book

The *Observer's B* ... *Aircraft* is the indi ... annual pocket guid... world's latest aeroplanes and most recent versions of established aircraft types. This, the thirty-second annual edition, embraces the latest fixed-wing and variable-geometry aeroplanes and rotorcraft of twenty-one countries. Its scope ranges from such regional airliners as the Dash 8, CN-235 and Brasilia, which will join the Saab-Fairchild 340 as major new débutantes of 1983, through important new derivative types, such as the Mirage IIING and F-20A Tigershark fighters, to the Su-25 *Frogfoot* and PA-48 Enforcer close air support aircraft; from new variants of such earlier-generation types as the DC-8, VC10 and KC-135 to such general aviation newcomers as the Cessna Caravan, Piper's Malibu and Mojave, and the Partenavia Spartacus. The ... ide ... types ... and illustrated; all data has been checked and revised as necessary, and most three-view silhouettes depicting types appearing in previous editions have been revised to reflect the latest changes introduced by their manufacturers.

About the Author

William Green, compiler of the *Observer's Book of Aircraft* for 32 years, is internationally known for many works of aviation reference. William Green entered aviation journalism during the early years of World War II, subsequently served with the RAF and resumed aviation writing in 1947. He is currently managing editor of one of the largest circulation European-based aviation journals, *Air International*, and co-editor of *Air Enthusiast* and the *RAF Yearbook*.

The New Observer's Book of

Aircraft

Compiled by

William Green

with silhouettes by

Dennis Punnett

Describing 142 aircraft
with 247 illustrations

1983 edition

Frederick Warne

Thirty-second Edition 1983

Library of Congress Catalog
Card No 57 4425

ISBN 0 7232 1640 1

Printed in Great Britain by
Butler & Tanner Ltd, Frome and London

INTRODUCTION TO THE 1983 EDITION

The *Observer's Book of Aircraft*, unlike most contemporary source books which devote themselves essentially to those aeroplanes most likely to be *seen*, is primarily concerned with the most recent aircraft types and variants currently in production, under test at the end of the preceding year, or scheduled to enter flight test during the year of the volume's currency. Reiteration of the *raison d'être* of this annual publication appears necessary owing to an increasing number of letters from readers querying omission of specific aircraft types remaining in production.

The *force majeure* of finding space for newcomers to the aviation scene dictates the discarding—with reluctance—each year some aircraft still retaining production significance. This, the 32nd yearly edition, is no exception, such important types as the Boeing 727, the Orion, the Mystère-Falcons, the HS 125 and 748, and the Harrier having, perforce, departed this year's volume. It must be stressed, however, that this annual is intended to be used in combination with the companion *Observer's Book of Airliners*, which embraces virtually *all* commercial transport aircraft to be seen today, and the larger *Observer's Directories* of military and civil aircraft, which, revised periodically, between them describe and illustrate the vast majority of aircraft currently in service.

Among the numerous civil débutantes to be found in the pages that follow are such new regional airliners as the Brasilia, the Dash 8 and the CN-235, which, together with the competitive Saab-Fairchild 340, will fly during 1983. In the general aviation sphere, newcomers range from the Cessna Caravan, the YS-11T Turbo-Panda and the Partenavia Spartacus utility aircraft to the Beechcraft Lightning and Piper's Malibu and Mojave tourers, and, further down the size and power scale, the Yak-55, Sheriff, Firefly and CAP X.

The military scene, too, is not without its new arrivals, such as the Mirage IIING and F-20A Tigershark multi-role fighters, the Su-25 *Frogfoot* and PA-48 Enforcer close air support aircraft, and instructional aircraft such as the IAR Triumf, the Ajeet Trainer and the Super Galeb. The Stratotanker, for decades the backbone of the USAF's flight refuelling force, reappears in its immensely more efficaceous KC-135R form and the Rockwell B-1 strategic bomber, after some time in limbo, has now been committed to production in B-1B form and is thus restored to the *Observer's Book*. Furthermore, earlier-generation airliners, the DC-8 and the VC10, make their reappearance after a lapse of many years in new versions, the latter now fulfilling a new role as a tanker aircraft.

WILLIAM GREEN

AERITALIA G.222

Country of Origin: Italy.
Type: General-purpose military transport.
Power Plant: Two 3,400 shp Fiat-built General Electric T64-GE-P4D turboprops.
Performance: (At 61,730 lb/28 000 kg) Max. speed, 336 mph (540 km/h) at 15,000 ft (4 575 m); long-range cruise, 273 mph (439 km/h) at 19,685 ft (6 000 m); max. initial climb, 1,705 ft/min (8,66 m/sec); service ceiling, 25,000 ft (7 620 m); range (max. payload), 852 mls (1 371 km), (53 troops), 1,497 mls (2 409 km), (max. fuel), 2,879 mls (4 633 km).
Weights: Empty equipped, 33,950 lb (15 400 kg); max. take-off, 61,730 lb (28 000 kg).
Accommodation: Normal flight crew of three and standard troop transport arrangement for 53 fully-equipped men or 42 paratroops. For aeromedical role 36 stretchers, two seated patients and four medical attendants may be accommodated.
Status: First and second prototypes flown 18 July 1970 and 22 July 1971, and first production G.222 flown 23 December 1975. Forty-four ordered by Italian *Aeronautica Militare*, three by Argentine Army, two by Somali Air Force, one by Dubai government, two by the Venezuelan Army and 20 by the Libyan Arab Air Force. Fifty-five G.222s delivered by the beginning of 1983, when production tempo was 0·9 monthly.
Notes: In addition to the standard G.222, a version powered by 4,860 shp Rolls-Royce Tyne RTy 20 turboprops, the G.222T (see 1982 edition), is being manufactured against the Libyan order. Of those ordered by the *Aeronautica Militare*, six are being delivered as G.222RMs (first flown on 11 October 1982) for navaid calibration. Illustrated above is a Venezuelan Army G.222, the drawing opposite depicting the G.222T.

AERITALIA G.222

Dimensions: Span, 94 ft 2 in (28,70 m); length, 74 ft 5½ in (22,70 m); height, 32 ft 1¾ in (9,80 m); wing area, 882·64 sq ft (82,00 m²).

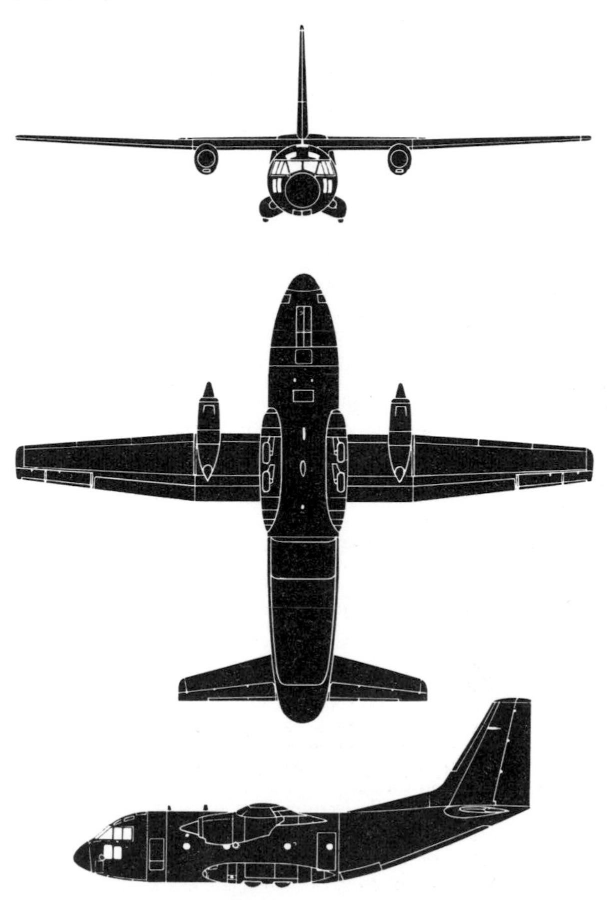

AERITALIA/PARTENAVIA AP68-300 SPARTACUS

Country of Origin: Italy.
Type: Light multi-role transport and utility aircraft.
Power Plant: Two 330 shp Allison B250-B17C turboprops.
Performance: (At 5,732 lb/2 600 kg) Max. cruising speed, 240 mph (386 km/h) at 15,000 ft (4 570 m); max. range speed, 190 mph (306 km/h) at 18,000 ft (5 485 m); initial climb, 2,034 ft/min (10,33 m/sec); service ceiling, 25,000 ft (7 620 m); range at max. cruise (with 1,600-lb/726-kg payload), 360 mls (576 km/h) at 10,000 ft (3 050 m), at econ cruise (with 1,032-lb/486-kg payload), 1,046 mls (1 683 km).
Weights: Empty, 3,245 lb (1 472 kg); max. take-off, 5,732 lb (2 600 kg).
Accommodation: Pilot and co-pilot/passenger in cockpit with three rows of paired individual seats in main cabin, the rearmost pair being replaceable by a bench-type seat for three.
Status: Prototype AP68TP flown on 11 September 1978, and first of five pre-series aircraft flown on 20 November 1981. Production version, the AP68-300, scheduled to enter flight test March 1983, with customer deliveries from initial series of 40 to commence second half of 1983. Eighteen on order at beginning of 1983.
Notes: Produced by Partenavia, a company of the Aeritalia General Aviation Division, the Spartacus has been evolved from the piston-engined P68 series, and differs from the pre-series AP68TP (illustrated above) primarily in replacing the 'all-flying' tailplane with a conventional unit of larger area and in having upturned wingtips. A pressurised derivative, the Pulsar, will enter flight test late 1984.

AERITALIA/PARTENAVIA AP68-300 SPARTACUS

Dimensions: Span, 39 ft 4½ in (12,00 m); length, 31 ft 9¾ in (9,70 m); height, 12 ft 0 in (3,65 m); wing area, 200·22 sq ft (18,60 m²).

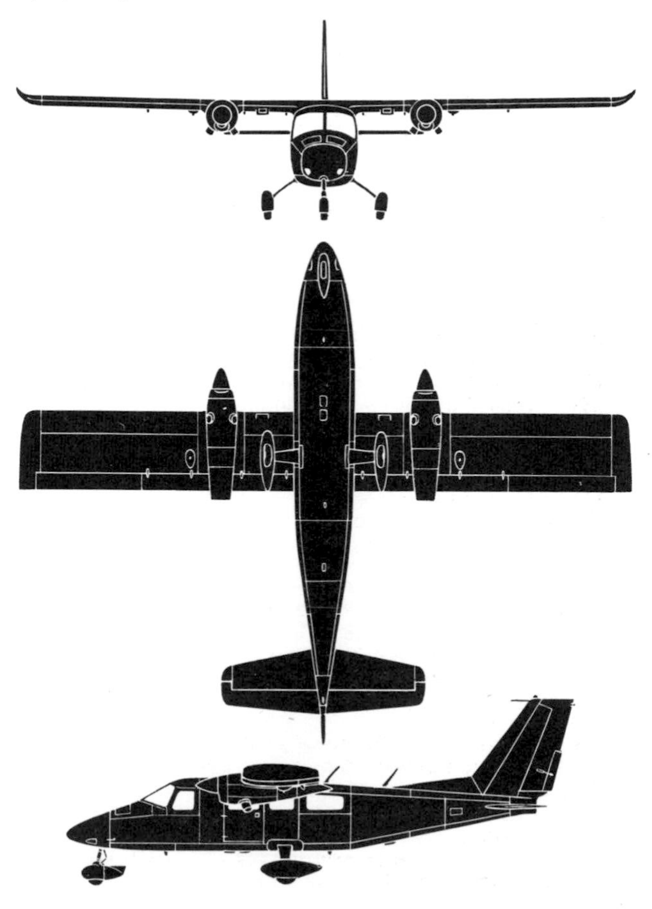

AERMACCHI MB-339A

Country of Origin: Italy.
Type: Tandem two-seat basic/advanced trainer.
Power Plant: One 4,000 lb st (1 814 kgp) Fiat-built Rolls-Royce Viper 632-43 turbojet.
Performance: Max. speed, 558 mph (898 km/h) at sea level, 508 mph (817 km/h) at 30,000 ft (9 145 m); initial climb, 6,595 ft/min (33,5 m/sec); time to 30,000 ft (9 145 m), 7·1 min; service ceiling, 48,000 ft (14 630 m); max. range (internal fuel with 10% reserves), 1 093 mls (1 760 km).
Weights: Empty, 6,780 lb (3 075 kg); normal loaded, 9,700 lb (4 400 kg); max. take-off, 13,000 lb (5 897 kg).
Armament: (Armament training and light strike) Max. of 4,000 lb (1 815kg) of ordnance may be distributed between six underwing stations when flown as single-seater.
Status: First of two prototypes flown 12 August 1976. First deliveries to Italian Air Force February 1981 against requirement for 100 aircraft, preceded by first export delivery (10 for Argentine Navy) commencing November 1980. Further export orders include 16 for the Peruvian Air Force and 12 (plus options) for Royal Malaysian Air Force. Indaer-Macchi at Callique, Peru, is to assemble a further 50 under licence in a programme to culminate in indigenous manufacture of wings, rear fuselage and tail assembly.
Notes: A single-seat light close air support version of the MB-339A, the MB-339K Veltro 2 (see 1981 edition) is currently available, having first flown on 30 May 1980, and is expected to be included among those assembled under licence in Peru by Indaer-Macchi. During 1982, the MB-339A was assigned secondary operational roles (eg, light strike and anti-helicopter missions) by the Italian Air Force.

AERMACCHI MB-339A

Dimensions: Span, 35 ft 7 in (10,86 m); length, 36 ft 0 in (10,97 m); height, 13 ft 1 in (3,99 m); wing area, 207·74 sq ft (19,30 m²).

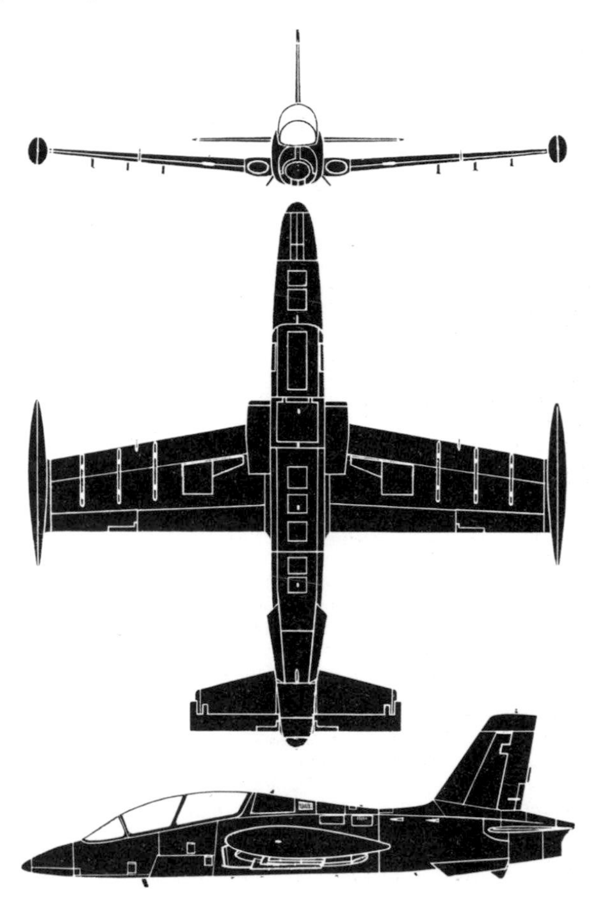

AÉROSPATIALE TB 30 EPSILON

Country of Origin: France.
Type: Tandem two-seat primary/basic trainer.
Power Plant: One 300 hp Avco Lycoming AEIO-540-L1B5-D six-cylinder horizontally-opposed engine.
Performance: (At 2,755 lb/1 250 kg) Max. speed, 238 mph (383 km/h) at 6,000 ft (1 830 m); max. cruise, 230 mph (370 km/h) at sea level; initial climb, 1,850 ft/min (9,4 m/sec); service ceiling, 20,010 ft (6 100 m); endurance (50% power), 3·75 hrs.
Weights: Empty equipped, 1,936 lb (878 kg); max. take-off, 2,755 lb (1 250 kg).
Status: First of two prototypes flown on 22 December 1979, and first 30 ordered against an *Armée de l'Air* requirement for 150 aircraft. Current schedules call for production deliveries to commence September 1983, with seven completed by the end of that year and service entry from June 1984. Production rate of four monthly is expected to be attained during 1984.
Notes: The Epsilon is scheduled to enter service at the *Armée de l'Air* college at Salon de Provence in June 1984, and, initially, will provide about 70 hours flying training after a few hours grading on the CAP 10 and before graduation to the Magister. However, it is claimed that the Epsilon could be utilised for 125 hours and lead straight to jet trainers in the category of the Alpha Jet and Hawk. The Epsilon is designed to emulate some of the flying characteristics of more advanced jet aircraft (eg, spinning and stalling), and an armed version is to be offered for export, with two or four wing stores stations for 440-660 lb (200-300 kg) of ordnance. The Epsilon is claimed to have approximately half the first cost of contemporary turboprop-powered trainers and its operating costs are appreciably lower.

AÉROSPATIALE TB 30 EPSILON

Dimensions: Span, 25 ft 11½ in (7,92 m); length, 24 ft 10½ in (7,59 m); height, 8 ft 8¾ in (2,66 m); wing area, 103·34 sq ft (9,60 m²).

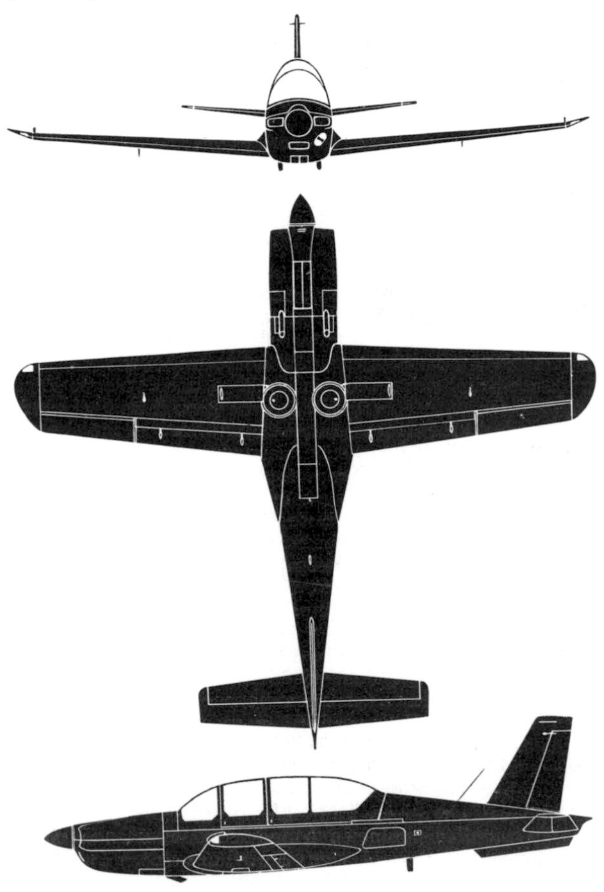

AIRBUS A300B4-200

Country of Origin: International consortium.
Type: Medium-haul commercial airliner.
Power Plant: Two 52,500 lb st (23 814 kgp) General Electric CF6-50C2 or 53,000 lb st (24 040 kgp) Pratt & Whitney JT9D-59A1 turbofans.
Performance: Max. cruising speed, 552 mph (889 km/h) at 31,000 ft (9 450 m); econ cruise, 535 mph (861 km/h) at 33,000 ft (10 060 m); long-range cruise, 530 mph (854 km/h); range (max. payload), 3,430 mls (5 520 km), (max. fuel and 61,600 lb/27 940 kg payload), 4,213 mls (6 780 km).
Weights: Operational empty, 194,900 lb (88 407 kg); max. take-off, 363,750 lb (165 000 kg).
Accommodation: Flight crew of three and various seating arrangements for 220-336 passengers in six-, seven-, eight- or nine-abreast seating.
Status: First A300B flown 28 October 1972, and first A300B4 on 26 December 1974. Production of both A300 and A310 (which see pages 16–17) running at five monthly at beginning of 1983, when orders for the A300 totalled 252 with 197 delivered.
Notes: The A300B is manufactured by a consortium of Aérospatiale (France), British Aerospace (UK), Deutsche Airbus (Federal Germany) and CASA (Spain). The A300B4 (described above) is a longer-range version of the B2, and the -200 differing from the -100 in having reinforced wings and fuselage, and strengthened undercarriage to cater for higher take-off weights. The A300B4-600, scheduled to fly late 1983, will feature the new, re-profiled rear fuselage of the A310 with an extension of the parallel portion of the fuselage, coupled with the new-generation engines offered with the A310, the launch customer for this variant being Saudia which will take delivery in the spring of 1984.

AIRBUS A300B4-200

Dimensions: Span, 147 ft 1¼ in (44,84 m); length, 175 ft 11 in (53,62 m); height, 54 ft 2 in (16,53 m); wing area, 2,799 sq ft (260,00 m²).

AIRBUS A310-200

Country of Origin: International consortium.
Type: Short/medium-haul commercial airliner.
Power Plant: Two 48,000 lb st (21 800 kgp) General Electric CF6-80A1 or Pratt & Whitney JT9D-7R4D1 turbofans, or 50,000 lb st (22 680 kgp) CF6-80A3 or JT9D-7R4E1 turbofans.
Performance: Max. cruising speed, 562 mph (904 km/h) at 33,000 ft (10 060 m); econ. cruise, 528 mph (850 km/h) at 37,000 ft (11 280 m); long-range cruise, 522 mph (840 km/h) at 39,000 ft (11 885 m); range (236 passengers), 2,995 mls (4 820 km), (max. payload), 1,440 mls (2 315 km).
Weights: Operational empty (typical), 175,863 lb (79 770 kg); max. take-off, 291,010 lb (132 000 kg), (option), 305,560 lb (138 600 kg).
Accommodation: Flight crew of two or three with single-class seating for 236 or 262 passengers eight abreast, or (typical mixed-class) 20 first-class six-abreast and 200 economy-class eight abreast.
Status: First A310 flown on 3 April 1982, with first scheduled for customer delivery (to Swissair) spring 1983. Total of 102 A310s ordered by beginning of 1983, when production rate (including A300B—see pages 14–15) was five monthly.
Notes: By comparison with the earlier A300B, the A310 has a new, higher aspect ratio wing, a shorter fuselage, a new, smaller tailplane and a new undercarriage, but retains a high degree of commonality with the preceding and larger aircraft. Like the A300B, the A310 is being built by a consortium of French, British, Federal German and Spanish companies, with Belgian (SONACA) and Dutch (Fokker) companies being associated with the programme. On offer, but not positively launched by the beginning of 1983, is the longer-range A310-300.

AIRBUS A310-200

Dimensions: Span, 144 ft 0 in (43,90 m); length, 153 ft 1 in (46,66 m); height, 51 ft 10 in (15,81 m); wing area, 2,357·3 sq ft (219,00 m²).

AIRCRAFT DESIGNS SHERIFF

Country of Origin: United Kingdom.
Type: Light cabin monoplane and trainer.
Power Plant: Two 160 hp Avco Lycoming O-320-D1A four-cylinder horizontally-opposed engines.
Performance: (Estimated) Max. speed, 173 mph (278 km/h); cruise (75% power), 158 mph (254 km/h), (60% power), 142 mph (229 km/h); initial climb, 1,300 ft/min (6,6 m/sec); max. range (no reserves), 668 mls (1 075 km).
Weights: (Estimated) Empty equipped, 1,885 lb (855 kg); max. take-off, 2,950 lb (1 338 kg).
Accommodation: (Tourer) four individual seats in side-by-side pairs, or (trainer) two seats side by side. Dual controls as standard.
Status: The first prototype Sheriff is scheduled to enter flight test in the second quarter of 1983, current planning envisaging production deliveries commencing late 1984.
Notes: The Sheriff, the prototype of which is being manufactured by Aircraft Designs (Bembridge), a Britten Aviation subsidiary company, places emphasis on lightweight, low initial cost and ease of maintenance, and is aimed primarily at the training and touring markets. Of essentially simple design and construction, the Sheriff was initially conceived by the late John Britten as an inexpensive and economical light twin-engined trainer, but its potential has since been developed to cater for the lower/middle range four-seat touring and air taxi market. At the end of 1982, consideration was being given to the possible manufacture of the Sheriff in Romania by the IAv Bucuresti organisation which would supply 'green' airframes to the parent company for equipping and finishing.

AIRCRAFT DESIGNS SHERIFF

Dimensions: Span, 33 ft 0 in (10,06 m); length, 22 ft 11 in (6,98 m); wing area, 150 sq ft (13,94 m²).

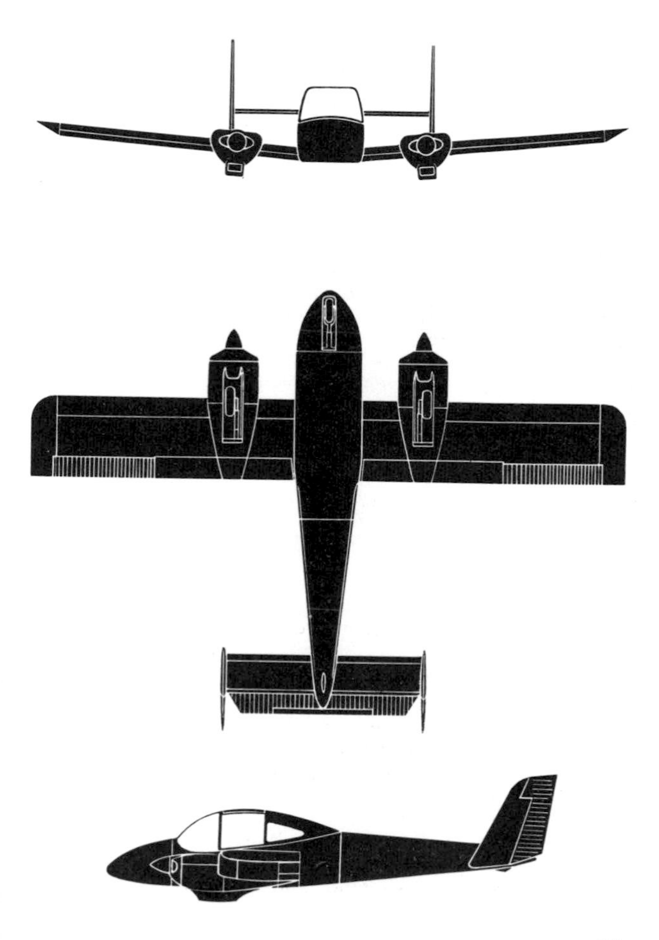

BEECHCRAFT 1900

Country of Origin: USA.
Type: Regional airliner and corporate transport.
Power Plant: Two 1,000 shp Pratt & Whitney (Canada) PT6A-65B turboprops.
Performance: Max. cruising speed, 303 mph (488 km/h) at 10,000 ft (3 050 m); long-range cruise, 250 mph (402 km/h) at 10,000 ft (3 050 m); initial climb, 2,280 ft/min (11,6 m/sec); service ceiling, 30,000 ft (9 150 m); max. range (19 passengers at cost econ cruise), 639 mls (1 028 km) at 10,000 ft (3 050 m), 977 mls (1 572 km) at 25,000 ft (7 620 m).
Weights: Operational empty (standard), 8,500 lb (3 856 kg); max. take-off, 15,245 lb (6 915 kg).
Accommodation: Flight crew of two and 19 passengers two-abreast with central aisle (with optional arrangement of 16 passengers two-abreast plus three abreast at rear of cabin). Optional rear cargo door to facilitate transportation of outsized loads.
Status: First prototype flown 3 September 1982, with second following in November. Third aircraft is being used in certification programme, the fourth being the first customer aircraft with delivery scheduled for September 1983 to coincide with certification. Current planning calls for a production rate of four monthly by early 1984.
Notes: Derived from the Super King Air, with which there is approximately 40 per cent commonality of component parts, the Beechcraft 1900 has been optimised for non-refuelling multi-stop operation. Aerodynamic features include so-called stabilons (horizontal fixed surfaces below and forward of the basic tail surfaces) to improve control in the pitch axis, and "tailets" projecting beneath the tailplane (increasing directional stability), plus vortex generators ahead of the wing (reducing stalling speed and drag).

BEECHCRAFT 1900

Dimensions: Span, 54 ft 6 in (16,61 m); length, 57 ft 10 in (17,63 m); height, 14 ft 10¾ in (4,53 m); wing area, 303 sq ft (28,15 m²).

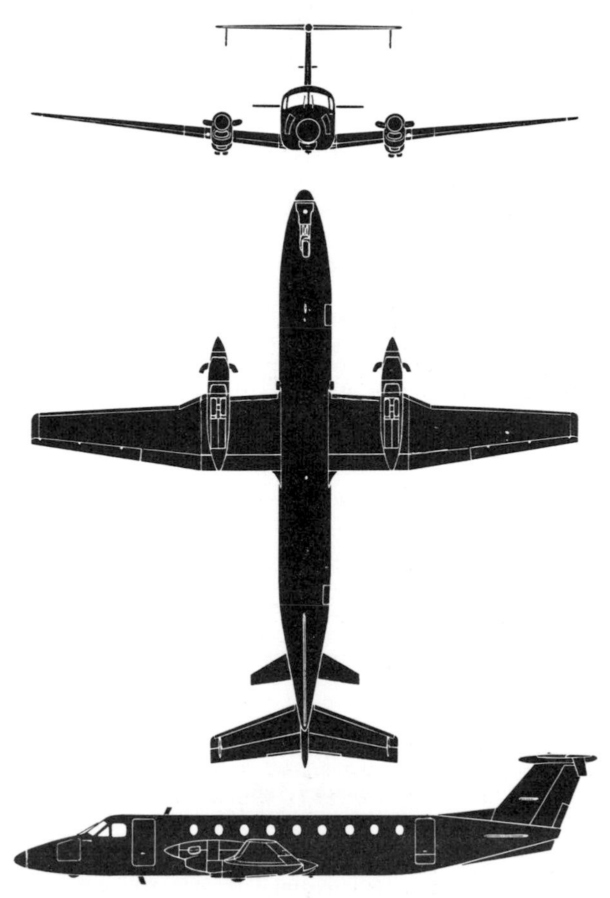

BEECHCRAFT LIGHTNING MODEL 38P

Country of Origin: USA.
Type: Light cabin monoplane.
Power Plant: One 865 shp Garrett TPE331-9, 850 shp Pratt & Whitney (Canada) PT6A-40, or 750 shp PT6A-130 turboprop.
Performance: Max. cruising speed (TPE331-9), 270 mph (435 km/h), (PT6A-40), 317 mph (510 km/h), (PT6A-130), 265 mph (426 km/h); max. altitude, 25,000 ft (7 620 m); range 1,150+ mls (1 850+ km).
Weights: No details available for publication.
Accommodation: Four individual seats in pairs, with fifth and sixth seats optional.
Status: Prototype Lightning flown on 14 June 1982, and production deliveries scheduled to commence late 1984 with engine options as listed above.
Notes: The Lightning Model 38P has been based upon the wing and fuselage structure of the twin piston-engined pressurised Baron Model 58P, and is being offered in three models differing primarily in power plant, Beech Aircraft having begun to accept delivery positions during September 1982. The Lightning possesses the advantage of utilising thoroughly proven structural components, the original Baron (Model 95-55) having first flown on 29 February 1960, and having since been manufactured continuously in progressively refined and more powerful versions, the Model 58 being introduced late 1969, and the Model 58P four years later. The first prototype is powered by the TPE331-9 engine and the second with a PT6A-40.

BEECHCRAFT LIGHTNING MODEL 38P

Dimensions: (Approximate) Span, 38 ft 0 in (11,58 m); length, 30 ft 0 in (9,14 m); height, 9 ft 0 in (2,74 m).

BEECHCRAFT T-34C (TURBINE MENTOR)

Country of Origin: USA.
Type: Tandem two-seat primary trainer.
Power Plant: One 400 shp Pratt & Whitney (Canada) PT6A-25 turboprop.
Performance: Max. speed, 246 mph (396 km/h) at 17,000 ft (5 180 m); cruising speed, 208 mph (335 km/h) at 1,000 ft (305 m), 233 mph (375 km/h) at 10,000 ft (3 050 m); initial climb, 1,480 ft/min (7,52 m/sec); time to 20,000 ft (6 095 m), 15 min; range, 492 mls (792 km) at 1,000 ft (305 m), 602 mls (970 m) at 10,000 ft (3 050 m), 815 mls (1 312 km) at 20,000 ft (6 095 m).
Weights: Empty, 2,940 lb (1 334 kg); max. take-off, 4,300 lb (1 950 kg).
Status: First of two YT-34Cs flown 21 September 1973, with first T-34C for US Navy following in August 1976. Total of 304 ordered for US Navy by beginning of 1983 (with deliveries scheduled for completion August 1985) against total requirement of approximately 450 aircraft.
Notes: Export versions of the basic T-34C are designated Turbine Mentor 34C and (with armament provisions) as the T-34C-1. The latter has been supplied to Argentina (15), Ecuador (23), Gabon (4), Indonesia (16), Morocco (12), Peru (6) and Uruguay (3). The T-34C-1 is suitable for both armament training and light counter-insurgency or close support missions, having four wing ordnance stations of 600 lb (272 kg) capacity inboard and 300 lb (136 kg) outboard, maximum combined load being 1,200 lb (544 kg). Six Turbine Mentor 34Cs have been supplied to Algeria's national pilot training school.

BEECHCRAFT T-34C (TURBINE MENTOR)

Dimensions: Span, 33 ft $4\frac{3}{4}$ in (10,18 m); length, 28 ft $8\frac{1}{2}$ in (8,75 m); height, 9 ft $10\frac{7}{8}$ in (3,02 m); wing area, 179·56 sq ft (16,68 m²).

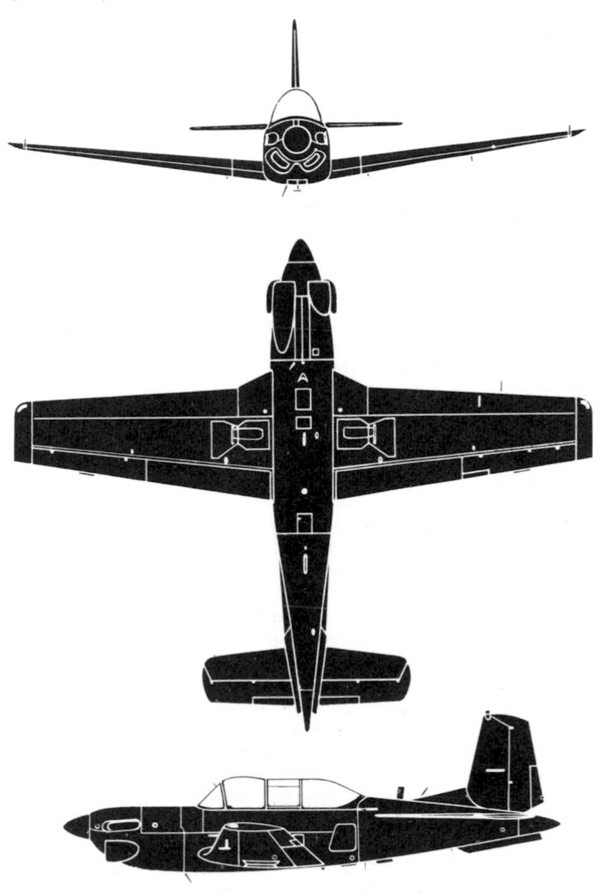

BOEING 737-200

Country of Origin: USA.
Type: Short-haul commercial airliner.
Power Plant: Two 16,000 lb st (7 258 kgp) Pratt & Whitney JT8D-17 turbofans.
Performance: Max. cruising speed, 564 mph (908 km/h) at 25,000 ft (7 620 m); econ cruise, 502 mph (808 km/h) at 33,000 ft (10 055 m); long-range cruise, 481 mph (775 km/h) at 35,000 ft (10 670 m); range (with max. fuel), 3,086 mls (4 967 km), (with max. payload), 1,750 mls (2 817 km).
Weights: Operational empty, 61,210 lb (27 764 kg); max. take-off, 117,000 lb (53 071 kg), (optional), 128,600 lb (58 333 kg).
Accommodation: Flight crew of two and up to 130 passengers in six-abreast seating, with optional arrangement for 115 passengers.
Status: First Model 737 flown 9 April 1967, with first deliveries (Lufthansa) following same year. Stretched 737-200 flown on 8 August 1967. Total of 1,046 on order (including 25 Model 737-300) at beginning of 1983, with 914 delivered. Ninety-eight Model 737-200s delivered during 1982, with production currently running at 10 monthly. Deliveries of Model 737-300 to commence November 1984.
Notes: Retaining 80 per cent commonality with the Model 737-200, the Model 737-300 to enter flight test March 1984 is a major new derivative of the basic design, with 20,000 lb st (9 072 kgp) CFM 56-3 turbofans and an 8 ft 8 in (2,64 m) fuselage stretch. The Model 737-300 is expected to offer a 25 per cent lower fuel burn per seat mile than the -200, and in addition to the basic max. take-off weight of 124,500 lb (56 473 kg) optional weights of 130,000 lb (58 968 kg) and 135,000 lb (61 236 kg) are being offered. The proposed -400 will be stretched a further 8 ft 4 in (2,54 m).

BOEING 737-200

Dimensions: Span, 93 ft 0 in (28,35 m); length, 100 ft 0 in (30,48 m); height, 37 ft 0 in (11,28 m); wing area, 980 sq ft (91,05 m²).

BOEING 747-300

Country of Origin: USA.
Type: Long-haul commercial ariliner.
Power Plant: Four 54,750 lb st (24 835 kgp) Pratt & Whitney JT9D-7R4G2 turbofans.
Performance: Max. cruising speed, 583 mph (939 km/h) at 35,000 ft (10 670 m); econ cruise, 564 mph (907 km/h) at 35,000 ft (10 670 m); long-range cruise, 558 mph (898 km/h); range (max. payload at econ cruise), 6,860 mls (11 040 km), (max. fuel at long-range cruise), 8,606 mls (13 850 km).
Weights: Operational empty, 389,875 lb (176 847 kg); max. take-off, 833,000 lb (377 850 kg).
Accommodation: Normal flight crew of three and up to 69 passengers six-abreast on upper deck, plus basic mixed-class arrangement for 410 passengers, or 415 passengers nine-abreast or 484 10-abreast in economy class seating.
Status: First Model 747-300 flown on 5 October 1982, with first customer delivery (Swissair) scheduled for March 1983. Total of 595 of all versions of the Model 747 ordered by the beginning of 1983, with 566 delivered and production running at two per month, 26 having been delivered during the course of 1982.
Notes: The Model 747-300 differs from the -200 primarily in having a 23-ft (7,0-m) lengthening of the upper deck affording a (typical) 10 per cent increase in total accommodation. The first Model 747-100 was flown on 9 February 1969, and the first Model 747-200 on 11 October 1970. Active studies are being made of several derivative versions, including introduction of a higher aspect ratio, 250-ft (76,20-m) wing of reduced sweep, and a range of fuselage sizes, with a 25-ft (7,62-m) stretch increasing passenger capacity to a maximum of 650. A new full-length upper deck cabin is also under consideration.

BOEING 747-300

Dimensions: Span, 195 ft 8 in (59,64 m); length, 231 ft 4 in (70,51 m); height, 63 ft 5 in (19,33 m); wing area, 5,685 sq ft (528,15 m²).

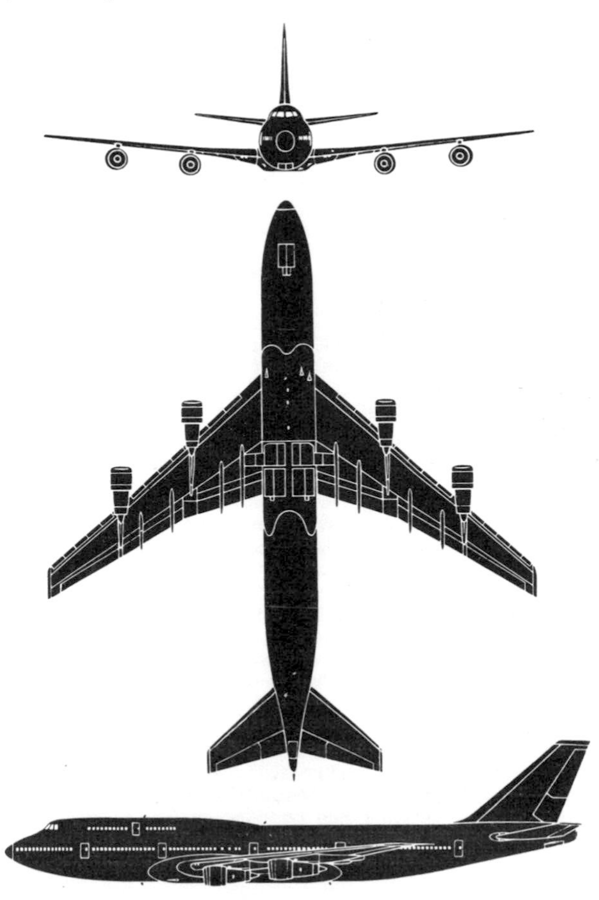

BOEING 757-200

Country of Origin: USA.
Type: Short/medium-haul commercial airliner.
Power Plant: Two 37,500 lb st (17 010 kgp) Rolls-Royce RB.211-535C, 38,200 lb st (17 327 kgp) Pratt & Whitney 2037 or 40,100 lb st (18 190 kgp) Rolls-Royce RB.211-535E4 turbofans.
Performance: (RB.211-535C engines) Max. cruising speed, 570 mph (917 km/h) at 30,000 ft (9 145 m); econ cruise, 528 mph (850 km/h) at 39,000 ft (11 885 m); range (max. payload), 2,210 mls (3 556 km) at econ cruise, (max. fuel), 5,343 mls (8 598 km) at long-range cruise.
Weights: Operational empty, 128,450 lb (58 265 kg); max. take-off (RB.211-535C engines), 220,000 lb (99 790 kg).
Accommodation: Flight crew of two (with provision for optional third crew member) and typical arrangement of 178 mixed class or 196 tourist class passengers, with max. single-class seating for 239 passengers.
Status: First Model 757 flown on 19 February 1982, with first customer deliveries (to Eastern) December 1982 and (British Airways) January 1983. Orders totalling 123 aircraft at beginning of 1983, of which approximately half to be powered by Pratt & Whitney engines with deliveries scheduled from end of 1984. Production rate of four aircraft monthly planned for 1983.
Notes: Two versions of the Model 757 are currently on offer, one with a max. take-off weight of 220,000 lb (99 790 kg) and the other for post-1984 delivery with a max. take-off weight of 240,000 lb (108 864 kg). The Model 757 is of narrowbody configuration and its wing has been optimised for short-haul routes. At the beginning of 1983, Boeing was engaged in studies of short-range, long-range and combi/convertible freighter versions of the aircraft.

BOEING 757-200

Dimensions: Span, 124 ft 6 in (37,82 m); length, 155 ft 3 in (47,47 m); height, 44 ft 6 in (13,56 m); wing area, 1,951 sq ft (181,25 m²).

BOEING 767-200

Country of Origin: USA.
Type: Medium-haul commercial airliner.
Power Plant: Two 48,000 lb st (21 773 kgp) Pratt & Whitney JT9D-7R4D or General Electric CF6-80A turbofans.
Performance: (JT9D-7R4D engines) Max. cruising speed, 556 mph (895 km/h) at 39,000 ft (11 890 m); econ cruise, 528 mph (850 km/h) at 39,000 ft (11 890 m); range (with max. payload and no reserves), 2,717 mls (4 373 km) at econ cruise, (max. fuel), 6,680 mls (10 750 km).
Weights: (JT9D-7R4D engines) Operational empty, 179,580 lb (81 457 kg); max. take-off, 300,000 lb (136 080 kg).
Accommodation: Flight crew of two (with optional three-crew arrangement) and typical mixed-class seating for 18 six-abreast and 193 seven-abreast with two aisles, with max. single-class seating for 290 passengers eight-abreast.
Status: First Model 767 (JT9D-7R4D engines) flown on 26 September 1981, (CF6-80A engines) 19 February 1982. First customer delivery (to United) on 18 August 1982, and 20 delivered (to seven customers) by beginning of 1983, when 177 were on order for 19 customers.
Notes: Three basic versions of the Model 767 were on offer at the beginning of 1983 with 300,000 lb (136 080 kg), 315,000 lb (142 884 kg) and 335,000 lb (151 956 kg) max. take-off weights, the last-mentioned version having 50,000 lb st (22 680 kg) CF6-80A2 engines and a max. volume payload range of 4,000 mls (6 437 km). Several variants are under consideration, including a stretched model—various stretches permitting increases of 20, 40 and 70 passengers—and a freighter, weights up to 360,000 lb (163 296 kg) being possible with the present wing and 55,000 lb st (24 948 kgp) engines. The longer-range 767-200ER has been ordered by Thai Airways and Ethiopian Airlines.

BOEING 767-200

Dimensions: Span, 156 ft 4 in (47,65 m); length, 159 ft 2 in (48,50 m); height, 52 ft 0 in (15,85 m); wing area, 3,050 sq ft (283,3 m²).

BOEING E-3 SENTRY

Country of Origin: USA.
Type: Airborne warning and control system aircraft.
Power Plant: Four 21,000 lb st (9 525 kgp) Pratt & Whitney TF33-PW-100A turbofans.
Performance: (At max. weight) Average cruising speed, 479 mph (771 km/h) at 28,900-40,100 ft (8 810-12 220 m); average loiter speed, 376 mph (605 km/h) at 29,000 ft (8 840 m); time on station (unrefuelled) at 1,150 mls (1 850 km) from base, 6 hrs, (with one refuelling), 14·4 hrs; ferry range, 5,034 mls (8 100 km) at 475 mph (764 km/h).
Weights: Empty, 170,277 lb (77 238 kg); normal loaded, 214,300 lb (97 206 kg); max. take-off, 325,000 lb (147 420 kg).
Accommodation: Operational crew of 17 comprising flight crew of four, systems maintenance team of four, a battle commander and an air defence team of eight.
Status: First of two (EC-137D) development aircraft flown 9 February 1972, two pre-production E-3As following in 1975. First 24 delivered to USAF as E-3As being modified to E-3B standards, and final 10 (including updated third test aircraft) are to be delivered as E-3Cs with completion of deliveries May 1984. Eighteen being delivered (in similar configuration to E-3C) to NATO as E-3As with deliveries to be completed June 1985, five similar aircraft being delivered to Saudi Arabia from August 1985. Thirty-two E-3s delivered by 1983.
Notes: Aircraft initially delivered to USAF as E-3As are now being fitted with JTIDS (Joint Tactical Information Distribution System), ECM-resistant voice communications, additional HF and UHF radios, austere maritime surveillance capability and more situation display consoles as E-3Bs. The E-3C will feature most E-3B modifications at the production stage, and the E-3As being delivered to NATO are of similar configuration to that of the USAF's E-3C.

BOEING E-3 SENTRY

Dimensions: Span, 145 ft 9 in (44,42 m); length, 152 ft 11 in (46,61 m); height, 42 ft 5 in (12,93 m); wing area, 2,892 sq ft (268,67 m²).

BOEING KC-135R STRATOTANKER

Country of Origin: USA.
Type: Flight refuelling tanker.
Power Plant: Four 22,000 lb st (9 979 kgp) General Electric/ SNECMA CFM56-2B-1 turbofans.
Performance: (Estimated) Typical cruising speed (max. load refuelling mission), 530 mph (853 km/h) at 35,000 ft (10 670 m); mission radius (to offload 30,000 lb/13 608 kg of fuel), 3,500 mls (5 630 km), (to offload 150,000 lb/68 040 kg of fuel), 1,250 mls (2 010 km).
Weights: Max. take-off, 322,500 lb (146 286 kg).
Accommodation: Flight crew of four comprising two pilots, a radio-navigator and a boom operator.
Status: First of initial batch of 10 KC-135Rs flown on 4 August 1982, with 10th scheduled for delivery to USAF in first quarter of 1985. The second batch is expected to comprise 22 KC-135Rs, current planning calling for modification of 300 early-model KC-135s to KC-135R standards through 1987. The French *Armée de l'Air* fleet of 11 KC-135Fs is to be similarly modified with initial deliveries mid-1985.
Notes: The KC-135R is a re-engined version of the KC-135A, the new engines, structural strengthening, an enlarged horizontal tail, a strengthened undercarriage and systems modifications being intended to extend the useful life of the Stratotanker to the year 2050. The modernisation programme increases max. fuel load from 189,702 lb (86 049 kg) to 203,288 lb (92 211 kg) and results in improved performance.

BOEING KC-135R STRATOTANKER

Dimensions: Span, 130 ft 10 in (39,88 m); length, 134 ft 6 in (40,99 m); height, 41 ft 8 in (12,69 m); wing area, 2,433 sq ft (226,03 m²).

BRITISH AEROSPACE 146-200

Country of Origin: United Kingdom.
Type: Short-haul regional airliner.
Power Plant: Four 6,968 lb st (3 160 kgp) Avco Lycoming ALF 502R-5 turbofans.
Performance: Max. cruising speed, 483 mph (778 km/h) at 26,000 ft (7 925 m); econ cruise, 441 mph (710 km/h) at 30,000 ft (9 145 m); long-range cruise, 436 mph (702 km/h) at 30,000 ft (9 145 m); range (max. payload), 1,232 mls (1 982 km) at econ cruise, (max. fuel), 1,440 mls (2 317 km), or (with optional fuel capacity), 1,727 mls (2 780 km).
Weights: Operational empty, 48,500 lb (22 000 kg); max. take-off, 89,500 lb (40 597 kg).
Accommodation: Flight crew of two and maximum seating (single-class) for 106 passengers six-abreast.
Status: First BAe 146-100 flown 3 September 1981, and first BAe 146-200 flown on 1 August 1982, with first customer deliveries of -100 (Dan-Air) early 1983, and -200 (Air Wisconsin) March 1983. Four -100s and 10 -200s ordered by beginning of 1983, and production rate of three aircraft monthly planned for late 1983 or early 1984.
Notes: The BAe 146 is currently being manufactured in -100 form with an 85 ft 10 in (26,16 m) fuselage for up to 82 passengers and -200 form (described). The BAe 146 is optimised for operation over stage lengths of the order of 150 miles (240 km) with unrefuelled multi-stop capability. Apart from fuselage length and capacity, the two versions of the BAe 146 are similar in all respects, but the uprated R-5 version of the ALF 502 turbofan is available for the longer -200 model which is featured by most initial orders. Various military versions of the BAe 146 have been considered, some featuring rear freight loading ramp, and two -100 models have been ordered for evaluation by the RAF.

BRITISH AEROSPACE 146-200

Dimensions: Span, 85 ft 5 in (26,34 m); length, 93 ft 8½ in (28,56 m); height, 28 ft 3 in (8,61 m); wing area, 832 sq ft (77,30 m²).

BRITISH AEROSPACE HAWK

Country of Origin: United Kingdom.
Type: Tandem two-seat basic/advanced trainer and light tactical aircraft.
Power Plant: One 5,200 lb st (2 360 kgp) Rolls-Royce Turboméca Adour 151, or (Srs 100) 5,700 lb st (2 585 kgp) Adour 861 turbofan.
Performance: (T Mk 1) Max. speed (clean aircraft with one crew member) 622 mph (1 000 km/h) or Mach 0·815 at sea level, 580 mph (933 km/h) or Mach 0·88 at 36,000 ft (10 970 m); max. climb, 11,833 ft/min (60,1 m/sec); tactical radius (with four 550-lb/250 kg bombs and two 130 Imp gal/590 l drop tanks), 680 mls (1 095 km) HI-LO-HI, 317 mls (510 km) LO-LO-LO.
Weights: Empty, 8,000 lb (3 629 kg); loaded (clean), 11,100 lb (5 040 kg); max. take-off (ground attack), 18,390 lb (8 342 kg).
Armament: (Srs 100) Five external ordnance stations for max. of 6,800 lb (3 100 kg) when flown as single-seater. Seventy-two Hawk T Mk 1s were being modified at beginning of 1983 to carry two AIM-9 Sidewinder AAMs to provide secondary air defence capability.
Status: Pre-series Hawk flown 21 August 1974, with first of 175 for RAF flown 19 May 1975. Export versions include Mk 51 (Finland, 50), Mk 52 (Kenya, 12), Mk 53 (Indonesia, 17) and Mk 54 (Zimbabwe, 8). The T-45A and T-45B are respectively carrier-capable and land-based developments of the Hawk for the US Navy, scheduled to fly in 1988 and 1987 respectively. The T-45 is being developed jointly with McDonnell Douglas, the US Navy having a requirement for 253 T-45As and 54 T-45Bs.

BRITISH AEROSPACE HAWK

Dimensions: Span, 30 ft 9¾ in (9,39 m); length, 38 ft 10⅔ in (11,85 m); height, 13 ft 1 in (4,00 m); wing area, 179·64 sq ft (16,69 m²).

BRITISH AEROSPACE JETSTREAM 31

Country of Origin: United Kingdom.
Type: Light corporate transport and regional airliner.
Power Plant: Two 900 shp Garrett AiResearch TPE 331-10 turboprops.
Performance: Max. cruising speed, 299 mph (482 km/h) at 20,000 ft (6 100 m); long-range cruise, 265 mph (426 km/h) at 25,000 ft (7 620 m); initial climb, 2,200 ft min (11,2 m/sec); max. range (with 19 passengers and IFR reserves), 737 mls (1 186 km), (with 12 passengers), 1,094 mls (1 760 km), (with nine passengers), 1,324 mls (2 130 km).
Weights: Empty equipped (including flight crew), 8,840 lb (4 010 kg); max. take-off, 14,550 lb (6 600 kg).
Accommodation: Two seats side by side on flight deck with basic corporate executive seating for eight passengers, with optional 12-seat executive shuttle arrangement, or up to 19 passengers three-abreast with offset aisle in high-density regional airline arrangement.
Status: First Jetstream 31 flown on 18 March 1982, following flight development aircraft (converted from a Series 1 airframe) flown on 28 March 1980. First customer delivery (Contactair of Stuttgart) made 15 December 1982. Eighteen aircraft are scheduled to be delivered during 1983, and 25 during 1984.
Notes: The Jetstream 31 is a derivative of the Handley Page H.P.137 Jetstream, the original prototype of which was flown on 18 August 1967. An inshore maritime patrol version, the Jetstream 31EZ (Economic Zone) was under development at the beginning of 1983, this having 360-deg scan search radar, a searchlight and a crew of five.

BRITISH AEROSPACE JETSTREAM 31

Dimensions: Span, 52 ft 0 in (15,85 m); length, 47 ft 2 in (14,37 m); height, 17 ft 6 in (5,37 m); wing area, 270 sq ft (25,08 m²).

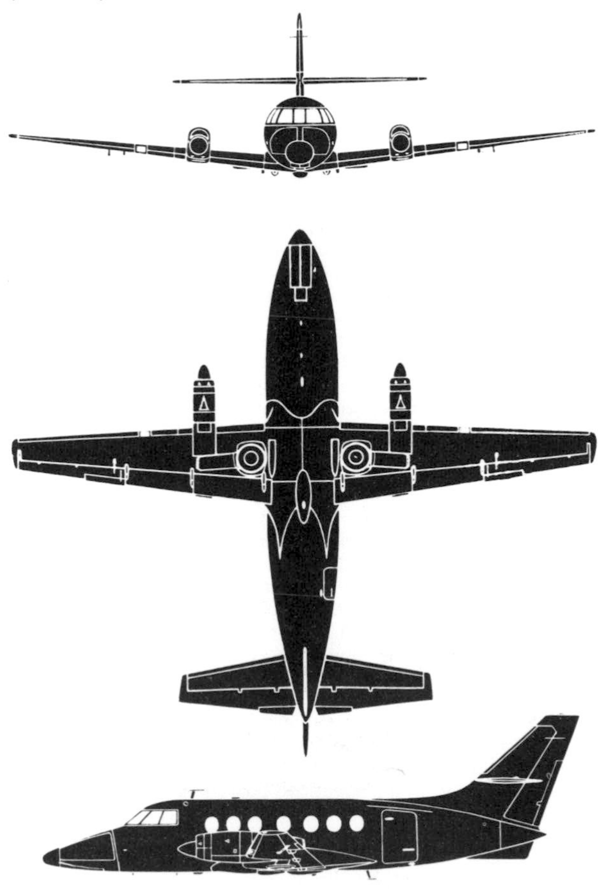

BRITISH AEROSPACE NIMROD MR MK 2

Country of Origin: United Kingdom.
Type: Long-range maritime patrol aircraft.
Power Plant: Four 12,160 lb st (5 515 kgp) Rolls-Royce RB.168-20 Spey Mk 250 turbofans.
Performance: Max. speed, 575 mph (926 km/h); max. transit speed, 547 mph (880 km/h); econ transit speed, 490 mph (787 km/h); typical ferry range, 5,180-5,755 mls (8 340-9 265 km); typical endurance, 12 hrs.
Weights: Max. take-off, 184,000 lb (83 460 kg); max. overload, 192,000 lb (87 090 kg).
Armament: Ventral weapons bay accommodating full range of ASW weapons (e.g, Stingray homing torpedoes, mines, depth charges). Provision for two underwing pylons on each side for total of four Aérospatiale AS 12 missiles.
Accommodation: Normal operating crew of 12 with two pilots and flight engineer on flight deck, and nine sensor operators and navigators in tactical compartment.
Status: Thirty-two Nimrod MR Mk 1s are being progressively brought up to MR Mk 2 standard in a programme scheduled to continue until mid-1984, the first MR Mk 2 production conversion having flown on 13 February 1979. The first of 46 Nimrod MR Mk 1s was flown on 28 June 1968, 11 of these being allocated to the AEW Mk 3 programme (see pages 46–47). In addition, three Nimrods have been supplied to the RAF as R Mk 1s for electronic reconnaissance duties.
Notes: The Nimrod MR Mk 2 possesses 60 times more computer power than the MR Mk 1 that it supplants in RAF service, and is equipped with the advanced Searchwater maritime radar, an AQS-901 acoustic system compatible with the Barra sonobuoy, and EWSM (Electronic Warfare Support Measures) in wingtip pods. During 1982, a number of Nimrod MR Mk 2s were fitted with air refuelling provisions.

BRITISH AEROSPACE NIMROD MR MK 2

Dimensions: Span, 114 ft 10 in (35,00 m); length, 126 ft 9 in (38,63 m); height, 29 ft 8½ in (9,01 m); wing area, 2,121 sq ft (197 05 m²).

BRITISH AEROSPACE NIMROD AEW MK 3

Country of Origin: United Kingdom.
Type: Airborne warning and control system aircraft.
Power Plant: Four 12,160 lb st (5 515 kgp) Rolls-Royce RB.168-20 Spey Mk 250 turbofans.
Performance: No details have been released for publication, but maximum and transit speeds are likely to be generally similar to those of the MR Mk 2 (see pages 44–45), and maximum endurance is in excess of 10 hours. The mission requirement calls for 6–7 hours on station at 29,000-35,000 ft (8 840-10 670 m) at approx 350 mph (563 km/h) at 750-1,000 miles (1 120-1 600 km) from base.
Weights: Max. take-off (approx), 190,000 lb (85 185 kg).
Accommodation: Flight crew of four and tactical team of six, latter comprising tactical air control officer, communications control officer, EWSM (Electronic Warfare Support Measures) operator and three air direction officers.
Status: Total of 11 Nimrod MR Mk 1 airframes being rebuilt to AEW Mk 3 standard of which fully representative prototype flew on 16 July 1980. The Nimrod AEW Mk 3 is scheduled to enter service with the RAF during 1983, by the end of which two will have been delivered, to become, in the following year, an element of the NATO airborne early warning mixed force.
Notes: The Nimrod AEW Mk 3 is equipped with Marconi mission system avionics with identical radar aerials mounted in nose and tail, these being synchronised and each sequentially sweeping through 180 deg in azimuth in order to provide uninterrupted coverage throughout the 360 deg of combined sweep. EWSM pods are located at the wingtips and weather radar in the starboard wing pinion tank. The Nimrod AEW Mk 3 is intended to provide complementary capability with the Boeing E-3A Sentries operated by the NATO combined force (excluding the UK).

BRITISH AEROSPACE NIMROD AEW MK 3

Dimensions: Span, 115 ft 1 in (35,08 m); length, 137 ft 8½ in (41,97 m); height, 35 ft 0 in (10,67 m); wing area, 2,121 sq ft (197,05 m²).

BRITISH AEROSPACE SEA HARRIER

Country of Origin: United Kingdom.
Type: Single-seat V/STOL shipboard milti-role fighter.
Power Plant: One 21,500 lb st (9 760 kgp) Rolls-Royce Pegasus 104 vectored-thrust turbofan.
Performance: Max. speed (clean aircraft), 720 mph (1 160 km/h) or Mach 0·95 at 1,000 ft (305 m), 607 mph (977 km/h) or Mach 0·92 at 36,000 ft (10 970 m), (with two AIM-9L AAMs and two Martel ASMs), 598 mph (962 km/h) or Mach 0·83 at sea level; combat radius (recce mission with two 100 Imp gal/455 l drop tanks), 518 mls (520 km); endurance (with two drop tanks for combat air patrol), 1·5 hrs at 115 mls/185 km from ship with three min combat.
Weights: Empty (approx), 13,000 lb (5 897 kg); normal loaded (STO), 21,700 lb (9 840 kg); max. take-off 25,600 lb (11 612 kg).
Armament: Provision for two 30-mm cannon plus two AIM-9L Sidewinder AAMs and up to 5,000 lb (2 268 kg) ordnance on five external stations.
Status: First Sea Harrier (built on production tooling) flown on 21 August 1978, with deliveries against initial 34 ordered for Royal Navy completed during 1982 when follow-on batch of 14 aircraft ordered. Six (FRS Mk 51) ordered for Indian Navy with completion of deliveries early 1983.
Notes: The Royal Navy's Sea Harrier FRS Mk 1 is a derivative of the RAF's Harrier GR Mk 3 (see 1982 edition) to operate from *Invincible*-class through-deck cruisers. Changes for the naval role include a new forward fuselage with raised cockpit, nose installation of Blue Fox intercept radar, new operational equipment and various changes to airframe and engine to suit maritime environment.

BRITISH AEROSPACE SEA HARRIER

Dimensions: Span, 25 ft 3 in (7,70 m); length, 47 ft 7 in (14,50 m); height, 12 ft 2 in (3,70 m); wing area, 201·1 sq ft (18,68 m²).

BRITISH AEROSPACE VC10 K MK 2

Country of Origin: United Kingdom.
Type: Flight refuelling tanker.
Power Plant: Four 21,800 lb st (9 888 kgp) Rolls-Royce Conway Mk 550B turbofans.
Performance: Max. cruising speed, 568 mph (914 km/h); long-range cruise, 425 mph (684 km/h) at 30,000 ft (9 150 m); initial climb, 3,050 ft/min (15,5 m/sec); approx. range, 3,900 mls (6 275 km).
Weights: Max. take-off (approx), 323,000 lb (146 510 kg).
Accommodation: Primary flight crew of four comprising pilot, co-pilot, navigator and flight engineer. Compartment for 18 personnel in aft-facing seating.
Status: First VC10 K Mk 2 (conversion of of VC10 Model 1101) flown 22 June 1982, and four similar conversions of the standard commercial VC10, plus four conversions of Super VC10 Model 1154s as K Mk 3s, for delivery to the RAF during course of 1983. Further VC10s will be converted as tankers as funding permits.
Notes: During 1958, the RAF initiated a programme to convert commercial VC10s as three-point tankers, this including both standard VC10s and the lengthened (171 ft 8 in/52,32 m) Super VC10s as K Mk 2s and K Mk 3s respectively, engine commonality being achieved and a common installation of fuselage fuel cells being adopted. After modification, the two versions of the tanker are assigned new Type numbers by British Aerospace, the K Mk 2s being Type 1112s and the K Mk 3s being Type 1164s. Fuel for refuelling operations is housed by five double-skinned cylindrical tanks, these being interconnected with the basic fuel system. Flight Refuelling Mk 32 hose-and-drogue pods are mounted beneath the wings and a Flight Refuelling Mk 17B HDU is installed in the rear fuselage to provide three-point refuelling.

BRITISH AEROSPACE VC10 K MK 2

Dimensions: Span, 146 ft 2 in (44,55 m); length (excluding refuelling probe), 158 ft 8 in (48,36 m); height, 39 ft 6 in (12,04 m); wing area, 2,932 sq ft (272,4 m²).

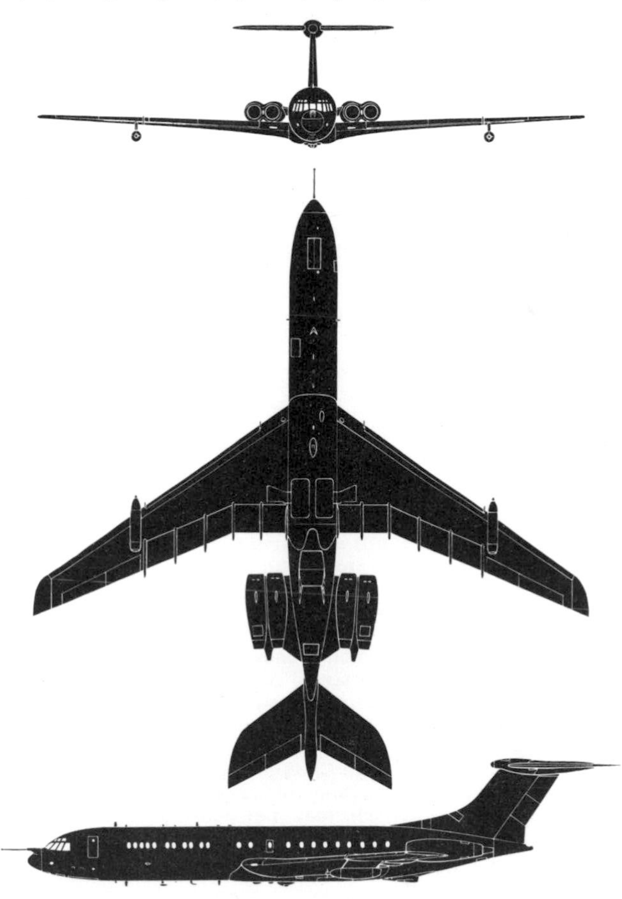

CANADAIR CL-601 CHALLENGER

Country of Origin: Canada.
Type: Light corporate transport.
Power Plant: Two 8,650 lb st (3 924 kgp) General Electric CF34-1A turbofans.
Performance: Max. cruising speed, 518 mph (834 km/h) or Mach 0·78; normal cruise, 497 mph (800 km/h) or Mach 0·75; range cruise, 463 mph (745 km/h) or Mach 0·7; time to max. operating altitude, 21 min; max. range (normal reserves), 4,030 mls (6 486 km).
Weights: Typical operational empty, 24,075 lb (10 920 kg); max. take-off, 41,650 lb (18 892 kg).
Accommodation: Flight crew of two with typical main cabin executive configurations for 8-11 passengers. Optional arrangements for 18 and 28 passengers for low- and high-density commuterline operation.
Status: Prototype CL-601 version of Challenger flown on 10 April 1982, with the first production example following on 17 September 1982. Orders for both CL-600 (transcontinental) and CL-601 (intercontinental) versions of the Challenger totalled some 170 at the beginning of 1983, including more than 45 of the latter variant, customer deliveries of which are scheduled to commence in the spring of 1983. Approximately 43 CL-600 Challengers delivered during 1982.
Notes: The CL-601 is basically similar to the CL-600 (see 1981 edition), but replaces 7,500 lb st (3 405 kgp) Lycoming ALF 502L-2 turbofans with CF34-1As and features winglets. The CL-600, which has a range of 3,224 mls (5 189 km) with standard tankage—which may be boosted to 3,685 mls (5 930 km) by optional fuselage tanks—is now described as the transcontinental model and the CL-601 is defined as the intercontinental model.

CANADAIR CL-601 CHALLENGER

Dimensions: Span, 64 ft 4 in (19,61 m); length, 68 ft 5 in (20,85 m); height, 20 ft 8 in (6,30 m); wing area (basic), 450 sq ft (41,82 m²).

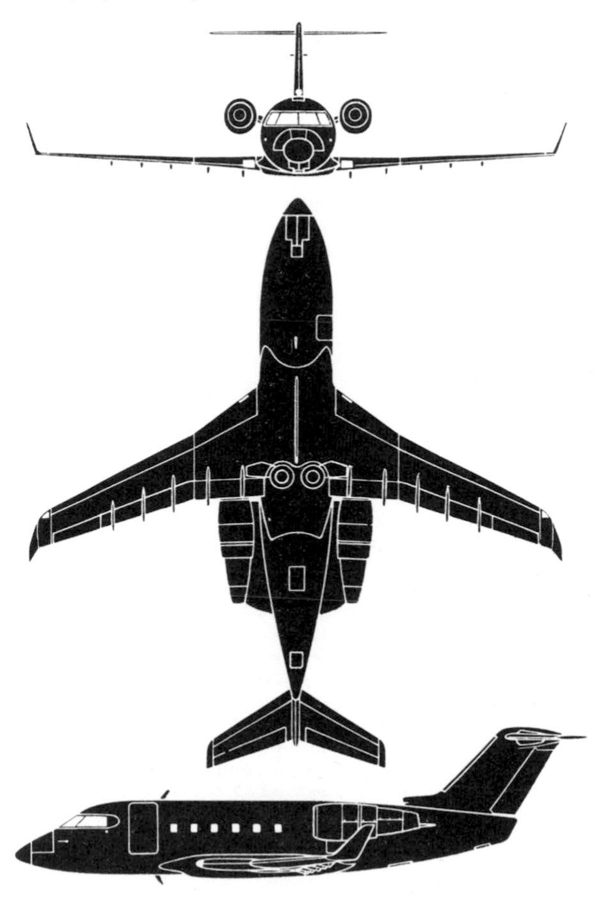

CAPRONI VIZZOLA C22-J2

Country of Origin: Italy.

Type: Side-by-side two-seat primary/basic trainer and (C22-J2/R) light tactical reconnaissance aircraft.

Power Plant: Two 286 lb st (130 kgp) Microturbo TRS 18 turbojets.

Performance: Max. speed, 322 mph (518 km/h); econ cruise (with wingtip tanks), 201 mph (324 km/h) at 9,845 ft (3 000 m); initial climb (without wingtop tanks), 1,811 ft/min (9,20 m/sec); time to 16,405 ft (5 000 m), 12 min; service ceiling, 25,100 ft (7 650 m); max. range (with wingtip tanks), 578 mls (930 km).

Weights: Empty equipped, 1,605 lb (728 kg); max. take-off, 2,767 lb (1 255 kg).

Status: The first prototype C22J was flown on 21 July 1980, and the second prototype, to definitive C22-J2 standards commenced flight test in October 1982. Two further prototypes built on production tooling scheduled to join the programme in January and April 1983, with first production deliveries planned for second quarter of 1984, a peak production tempo of five aircraft monthly being currently envisaged.

Notes: The C22-J2 embodies a number of changes by comparison with the original C22J (see 1982 edition), uprated engines having been installed, internal fuel capacity has been increased, the wingspan has been reduced and provision has been made for wingtip fuel tanks and wing hardpoints for two 275-lb (125-kg) capacity NATO pylons. To be manufactured by Aeronautica Caproni Vizzola, a component of the Agusta group, the C22-J2 is being offered as a low-cost reconnaissance system with a multi-spectral television system. In reconnaissance and tactical intelligence gathering form, it is capable of being launched from a mobile platform with rocket assistance and recovered by an arrester system.

CAPRONI VIZZOLA C22-J2

Dimensions: Span, 30 ft 2¼ in (9,20 m); length, 20 ft 6½ in (6,26 m); height, 6 ft 2 in (1,88 m); wing area, 82·35 sq ft (7,65 m²).

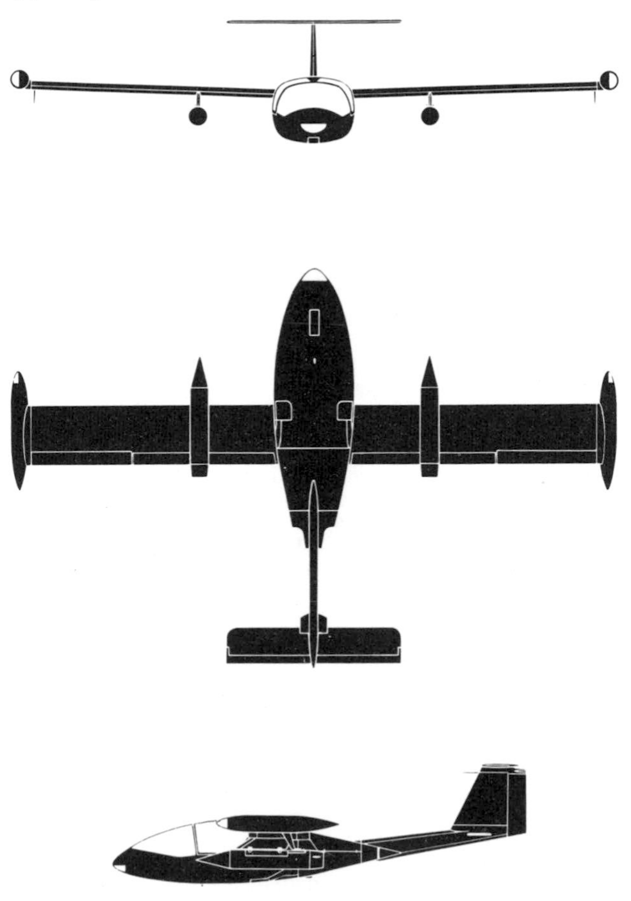

CASA C-101 AVIOJET

Country of Origin: Spain.
Type: Tandem two-seat basic/advanced trainer and light tactical support aircraft.
Power Plant: One (C-101EB) 3,500 lb st (1 588 kgp) Garrett AiResearch TFE 731-2-2J or (C-101BB) 3,700 lb st (1 678 kgp) TFE 731-3-1J turbofan.
Performance: (C-101BB) Max. speed (clean aircraft), 432 mph (695 km/h) at sea level, 495 mph (797 km/h) at 25,000 ft (7 620 m); initial climb, 3,800 ft/min (19,3 m/sec); time to 25,000 ft (7 620 m), 8·5 min; tactical radius (with four 550-lb/250-kg bombs and 30-mm cannon on internal fuel with 10 min reserves), 207 mls (333 km) LO-LO-LO interdiction.
Weights: (C-101BB) Empty equipped, 7,496 lb (3 400 kg); max. take-off, 12,346 lb (5 600 kg).
Armament: (C-101BB) One 30-mm cannon or two 12,7-mm machine guns plus max. of 3,307 lb (1 500 kg) of ordnance distributed between six wing stations.
Status: First of four prototypes flown 29 June 1977, with deliveries to Spanish Air Force against orders for 88 C-101EB trainers commencing March 1980, and some 60 delivered by beginning of 1983. First C-101BB delivered to Chile in August 1981 as T-36. Chilean requirement for 60 aircraft of which first four delivered complete with further aircraft being assembled in Chile by INDAER with progressive Chilean manufacturing participation. Proportion of Chilean production to comprise single-seat C-101CC dedicated attack model.
Notes: The C-101EB pilot trainer and dual-role C-101BB are similar apart from engine power and armament provisions. Versions of the Aviojet with more advanced equipment (eg, head-up display and low-level Doppler navigation system) were under development at beginning of 1983.

CASA C-101 AVIOJET

Dimensions: Span, 34 ft 9$\frac{3}{8}$ in (10,60 m); length, 40 ft 2$\frac{1}{4}$ in (12,25 m); height, 13 ft 11 in (4,25 m); wing area, 215·3 sq ft (20,00 m²).

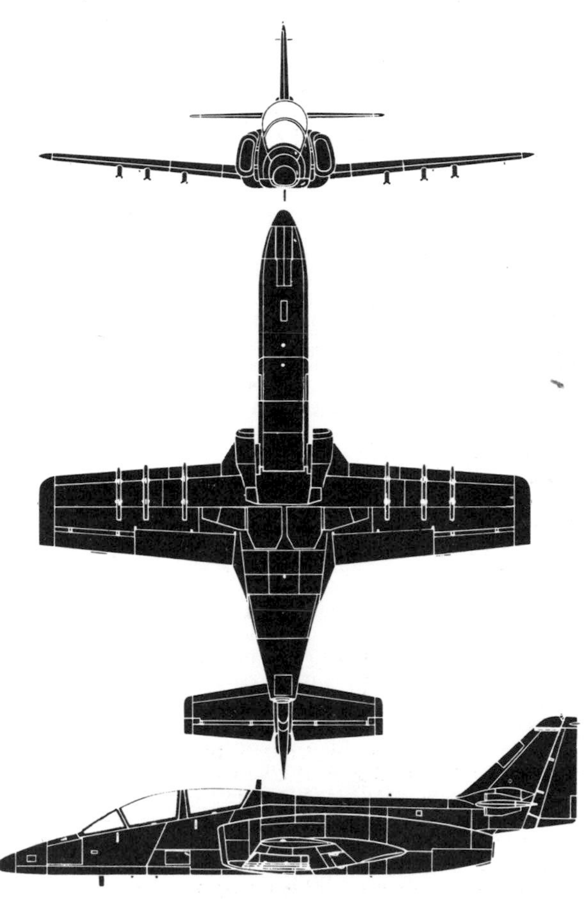

CASA-NURTANIO (AIRTEC) CN-235

Countries of Origin: Spain and Indonesia.
Type: Regional airliner, freighter and mixed passenger/freight transport.
Power Plant: Two 1,700 shp General Electric CT7-7 turboprops.
Performance: Max. cruising speed, 282 mph (454 km/h) at 20,000 ft (6 095 m); initial climb, 1,780 ft/min (9,04 m/sec); service ceiling, 28,800 ft (8 780 m); range (38 passengers at max. cruise with normal reserves), 690 mls (1 110 km), (34 passengers), 920 mls (1 480 km).
Weights: Max. zero fuel, 26,014 lb (11 800 kg); max. take-off, 28,658 lb (13 000 kg).
Accommodation: Flight crew of two and standard seating arrangement for 34 or 38 passengers four-abreast with central aisle, or all-cargo version with rear-loading ramp for up to 9,920 lb (4 500 kg) freight.
Status: Prototypes scheduled to fly simultaneously in Spain and Indonesia in October 1983, with certification mid-1984, and first customer deliveries November 1984. Total of 104 firm orders (including 32 for the Indonesian Air Force) recorded by beginning of 1983.
Notes: The CN-235 is being developed jointly by CASA of Spain and Nurtanio of Indonesia, and is being built on assembly lines in both countries without duplication of component manufacture. The CN-235 has been designed to meet both civil and military requirements, the latter including maritime surveillance, electronic intelligence, paratroop transportation and aeromedical evacuation. A stretched derivative of the CN-235 for 59 passengers is currently planned.

CASA-NURTANIO (AIRTEC) CN-235

Dimensions: Span, 84 ft 7½ in (25,80 m); length, 69 ft 10⅓ in (21,30 m); height, 25 ft 10⅞ in (7,90 m).

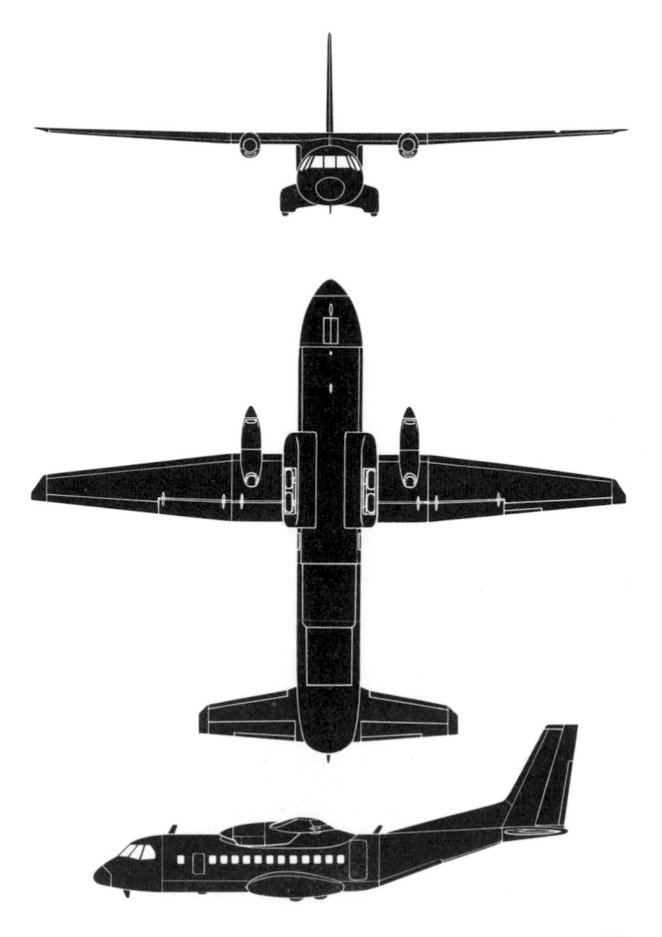

CESSNA CARAVAN

Country of Origin: USA.
Type: Light utility transport.
Power Plant: One 600 shp Pratt & Whitney (Canada) PT6A-114 turboprop.
Performance: Max. continuous cruise (at 6,700 lb/3 039 kg), 214 mph (344 km/h); range, 1,150 mls (1 850 km); initial climb, 1,500+ ft/min (7,62 m/sec).
Weights: Max. take-off, 6,700 lb (3 039 kg).
Accommodation: Pilot and up to 14 passengers in a combination of two- and three-abreast seating with aisle between seats
Status: First prototype Caravan flown on 9 December 1982, with certification scheduled for 1984, and full production commencing early 1985.
Notes: Claimed to be the first all-new single-engined turboprop-powered aircraft designed for the utility role, the Caravan is to be offered both in skiplane configuration and with Wipline amphibious floats, having a gross weight of 7,050 lb (3 198 kg) with the latter. An airstair door is provided for passengers in the starboard rear of the cabin, and baggage and cargo are loaded through a large two-piece door in the portside. The cabin can accommodate such loads as 10 55 US gal (208 1) fuel drums or two D-size cargo containers, and has a total volume of 340 cu ft (9,63 m³), including the baggage area. Optional installations will enable the Caravan to be used for aerial firefighting, agricultural spraying, aerial photography, supply dropping and aeromedical applications Seventy per cent of the wing span is occupied by flaps and the prototype also features a roll control system including roll spoilers activated with the ailerons.

CESSNA CARAVAN

Dimensions: Span, 51 ft 8 in (15,75 m); length, 37 ft 7 in (11,45 m); height, 14 ft 2 in (4,32 m).

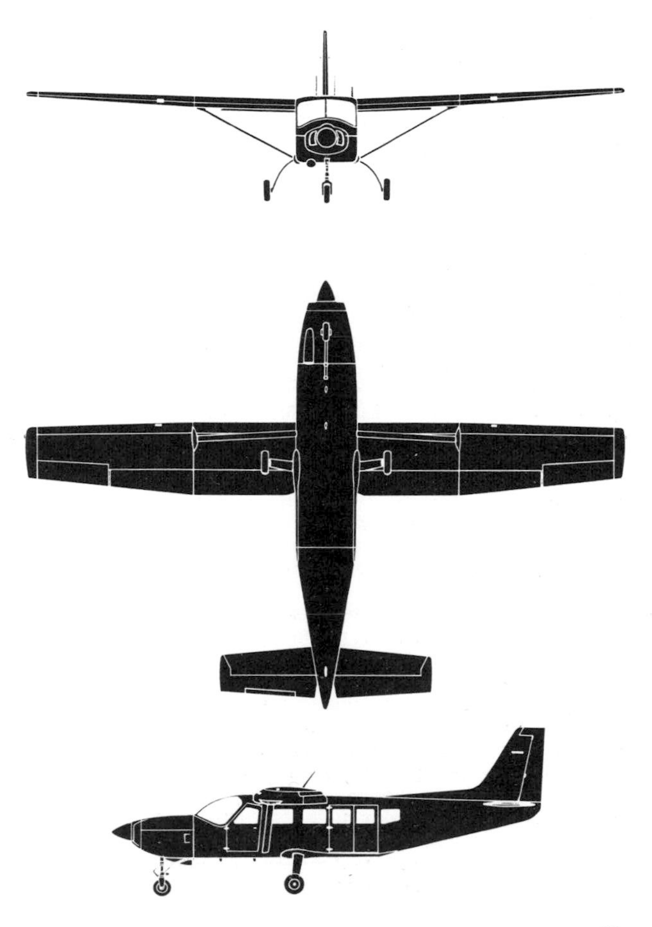

CESSNA CITATION III

Country of Origin: USA.

Type: Light corporate executive transport.

Power Plant: Two 3,650 lb st (1 656 kgp) Garrett AiResearch TFE731-3B-100S turbofans.

Performance: Max. cruising speed, 543 mph (874 km/h) at 33,000 ft (10 060 m), 528 mph (850 km/h) at 41,000 ft (12 500 m); max. initial climb, 4,475 ft/min (22,73 m/sec); time to 43,000 ft (13 105 m), 25 min; max. ceiling, 51,000 ft (15 545 m); max. range (six passengers and 45 min reserves), 2,994 mls (4 818 km).

Weights: Empty equipped, 10,951 lb (4 967 kg); max. take-off, 20,000 lb (9 072 kg).

Accommodation: Normal flight crew of two on flight deck and standard main cabin arrangement for six passengers in individual seats, with optional arrangements for seven to ten passengers.

Status: Two prototypes flown on 30 May 1979 and 2 May 1980 respectively, with first production aircraft flown October 1982 and first customer delivery following in December. Some 150 ordered by beginning of 1983, with 15 to be delivered by 30 September of that year when production will have attained four monthly, rising to eight monthly by late 1984.

Notes: The Citation III possesses no commonality with the Citation II (see 1982 edition) despite its name. It is claimed to offer an appreciably shorter balanced field length at 4,350 ft (1 326 m) than other aircraft in its class, permitting use of smaller airfields. Plans to offer an extended range version with an additional fuel cell aft of the rear cabin pressure bulkhead have been discarded.

CESSNA CITATION III

Dimensions: Span, 53 ft $3\frac{1}{2}$ in (16,30 m); length, 55 ft 6 in (16,90 m); height, 17 ft $3\frac{1}{2}$ in (5,30 m); wing area, 312 sq ft (29,00 m²).

DASSAULT-BREGUET ATLANTIC G2 (ATL2)

Country of Origin: France.
Type: Long-range maritime patrol aircraft.
Power Plant: Two 5,665 shp Rolls-Royce/SNECMA Tyne RTy 20 Mk 21 turboprops.
Performance: Max. speed, 368 mph (593 km/h) at sea level; normal cruise, 345 mph (556 km/h) at 25,000 ft (7 620 m); typical patrol speed, 196 mph (315 km/h); initial climb, 2,000 ft/min (10,1 m/sec); service ceiling, 30,000 ft (9 100 m); typical mission, 8 hrs patrol at 690 mls (1 110 km) from base at 2,000-3,000 ft (610-915 m); max. range, 5,590 mls (9 000 km).
Weights: Empty equipped, 56,217 lb (25 500 kg); normal loaded weight, 97,885 lb (44 400 kg); max. take-off, 101,850 lb (46 200 kg).
Accommodation: Normal flight crew of 12, comprising two pilots, flight engineer, forward observer, radio navigator, ESM/ECM/MAD operator, radar operator, tactical co-ordinator, two acoustic operators and two aft observers.
Armament: Up to eight Mk 46 homing torpedoes, nine 550-lb (250-kg) bombs or 12 depth charges, plus two AM 39 Exocet ASMs in forward weapons bay. Four wing stations with combined capacity of 7,715 lb (3 500 kg).
Status: First of two prototypes (converted from ATL1s) flown 8 May 1981, with second following on 26 March 1982, decision to manufacture 42 for France's *Aéronavale* announced June 1982, with production launch in 1984, and first deliveries scheduled for 1988.
Notes: The Atlantic G2 (*Génération* 2), also referred to as the ATL2, is a modernised version of the Atlantic G1 (now referred to as the ATL1), production of which terminated in 1973 after completion of 87 series aircraft. By comparison with the ATL1, the ATL2 has upgraded systems and embodies refined structural techniques.

DASSAULT-BREGUET ATLANTIC G2 (ATL2)

Dimensions: Span, 122 ft 7 in (37,36 m); length, 107 ft $0\frac{1}{4}$ in (36,62 m); height, 37 ft $1\frac{1}{4}$ in (11,31 m); wing area, 1,295·3 sq ft (120,34 m²).

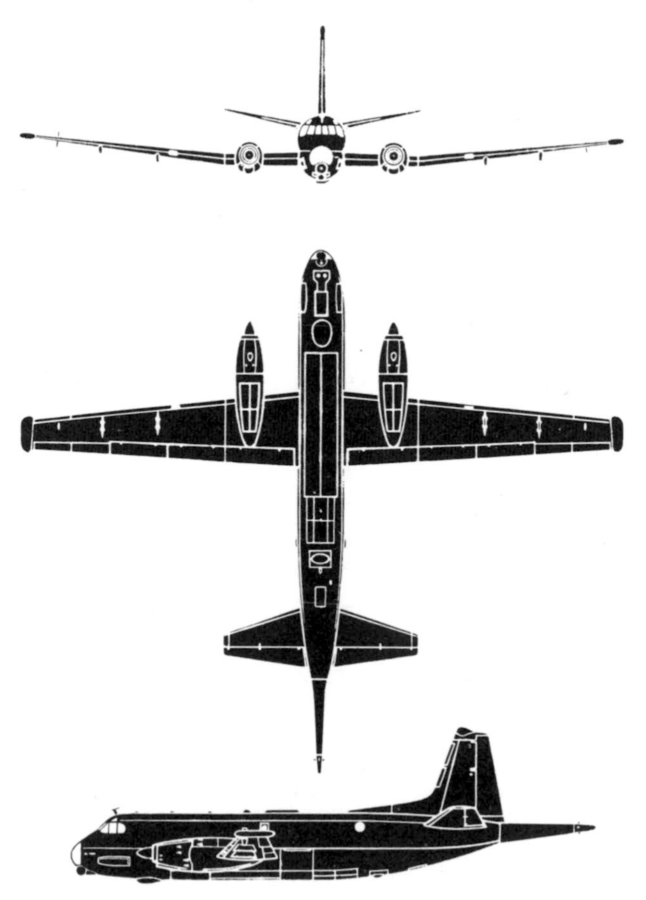

DASSAULT-BREGUET MIRAGE IIING

Country of Origin: France.
Type: Single-seat multi-role fighter.
Power Plant: One 11,023 lb st (5 000 kgp) dry and 15,873 lb st (7 200 kgp) reheat SNECMA 9K50 turbojet.
Performance: Max. speed (clean condition), 1,460 mph (2 350 km/h) or Mach 2·2 at 39,370 ft (12 000 m), 860 mph (1 385 km/h) or Mach 1.13 at sea level; combat radius (with 1,764-lb/800-kg external ordnance load), 745 mls (1 200 km) HI-LO-HI, 404 mls (650 km) LO-LO-LO.
Weights: Approx empty equipped, 15,873 lb (7 200 kg); max. take-off, 31,000 lb (14 060 kg).
Armament: Two 30-mm DEFA cannon, plus (intercept) two Matra Magic and one Matra R.530 AAMs, or (attack) up to 8,818 lb (4 000 kg) of external ordnance.
Status: Prototype Mirage IIING flown on 21 December 1982.
Notes: The Mirage IIING (*Nouvelle Génération*) has been developed as an export fighter to compete with the IAI Kfir-C2 (see 1982 edition) and Northrop F-20A Tigershark (see pages 152-3). It is based on the Mirage IIIE airframe to which have been added leading-edge wing root extensions and intake-mounted canard surfaces, and utilises the Atar 9K50 turbojet of the Mirage 50 and F1. It incorporates a fly-by-wire flight control system derived from that of the Mirage 2000 and the multi-purpose Cyrano IV radar of the Mirage F1. All systems have been modernised.

DASSAULT-BREGUET MIRAGE IIING

Dimensions: Span, 27 ft 0 in (8,22 m); length, 51 ft $0\frac{1}{2}$ in (15,56 m); height, 14 ft 9 in (4,50 m).

DASSAULT-BREGUET MIRAGE F1

Country of Origin: France.
Type: Single-seat multi-role fighter.
Power Plant: One 11,023 lb st (5 000 kgp) dry and 15,873 lb st (7 200 kgp) reheat SNECMA Atar 9K50 turbojet.
Performance: (F1C) Max. speed (clean aircraft), 914 mph (1 470 km/h) or Mach 1·2 at sea level, 1,450 mph (2 555 km/h) or Mach 2·2 at 39,370 ft (12 000 m); initial climb, 41,930 ft/min (213 m/sec); service ceiling, 65,600 ft (20 000 m); tactical radius (with two drop tanks and 4,410 lb/2 000 kg bombs), 670 mls (1 078 km).
Weights: Empty, 16,314 lb (7 400 kg); normal loaded, 24,030 lb (10 900 kg); max. take-off, 32,850 lb (14 900 kg).
Armament: Two 30-mm DEFA 553 cannon and (intercept) one-three Matra 550 Magic plus two AIM-9 AAMs, or (close support) up to 8,818 lb (4 000 kg) of external ordnance.
Status: First of four prototypes flown 23 December 1966, and first production aircraft flown 15 February 1973, with some 570 delivered by beginning of 1983 against orders for 678 for 11 countries and production continuing at seven monthly.
Notes: Current production models for the *Armée de l'Air* consist of the F1C-200 with fixed flight refuelling probe (illustrated), the tactical recce F1CR and the two-seat F1B conversion trainer. Whereas the F1C is an optimised air-air version, the export F1A and F1E are optimised for the air-ground role. Foreign orders for the F1 are Ecuador (18), Iraq (89), Jordan (36), Qatar (14), Kuwait (20), Libya (38), Morocco (50), Greece (40), South Africa (48) and Spain (73). The 500th F1 was delivered in April 1982, at which time the *Armée de l'Air* had received 167 aircraft, including the two prototype F1CRs, 60 production examples of which have been ordered.

DASSAULT-BREGUET MIRAGE F1

Dimensions: Span, 27 ft 6¾ in (8,40 m); length, 49 ft 2½ in (15,00 m); height, 14 ft 9 in (4,50 m); wing area, 269·1 sq ft (25,00 m²).

DASSAULT-BREGUET MIRAGE 2000

Country of Origin: France.
Type: Single-seat multi-role fighter.
Power Plant: One 12,345 lb st (5 600 kgp) dry and 19,840 lb st (9 000 kgp) reheat SNECMA M53-5, or (from 1985) 14,460 lb st (6 560 kgp) dry and 21,385 lb st (9 700 kgp) reheat M53-P2 turbofan.
Performance: Max. speed (clean aircraft), 915 mph (1 472 km/h) or Mach 1·2 at sea level, 1,550 mph (2 495 km/h) or Mach 2·35 (short endurance dash) above 36,090 ft (11 000 m); max. climb, 49,000 ft/min (249 m/sec); service ceiling, 59,055 ft (18 000 m); combat radius (intercept mission with two drop tanks and four AAMs), 435 mls (700 km).
Weights: Loaded (clean), 20,944 lb (9 500 kg); max. take-off, 33,070 lb (15 000 kg).
Armament: Two 30-mm DEFA 554 cannon and (air superiority) two Matra 550 Magic and two Matra Super 530D AAMs, or (close support) up to 13,227 lb (6 000 kg) of ordnance distributed between nine stations.
Status: First of five prototypes (four single-seat and one two-seat) flown 10 March 1978, with first production Mirage 2000C flown 20 November 1982, at which time 108 funded against anticipated *Armée de l'Air* requirement for 400 in three main versions. Orders placed during 1982 by Egypt (20), Peru (20) India (40), and first two-seat Mirage 2000N (*Nucléaire*) scheduled to enter flight test 1983, funding for 15 of this version (for 1988 delivery) included in 1983 budget.
Notes: The Mirage 2000 is to be adopted by the *Armée de l'Air* in single-seat air superiority and attack, and two-seat conversion training and low-level penetration versions, deliveries being scheduled to commence during 1983 with service entry following during 1984.

DASSAULT-BREGUET MIRAGE 2000

Dimensions: Span, 29 ft 6 in (9,00 m); length, 47 ft $6\frac{7}{8}$ in (14,50 m); wing area, 441·3 sq ft (41,00 m²).

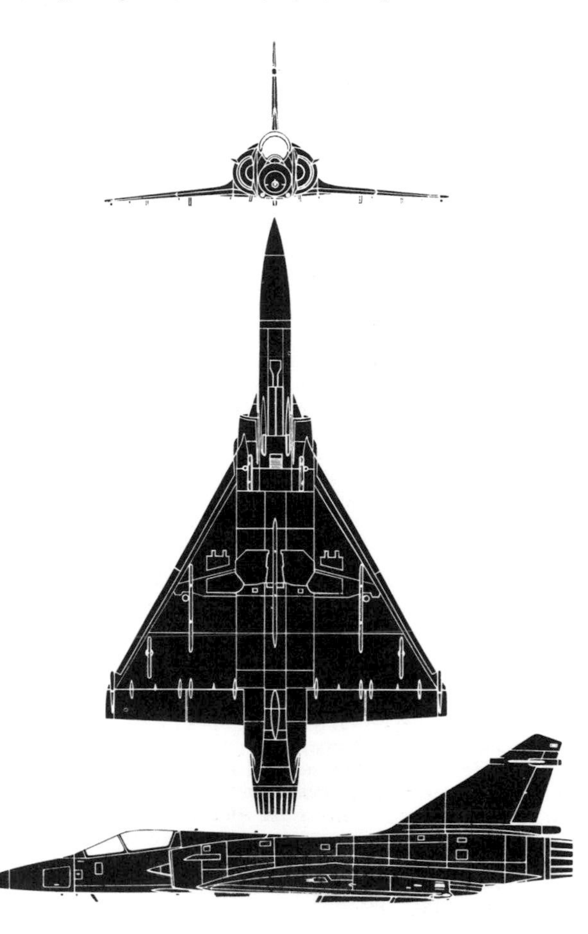

DASSAULT-BREGUET/DORNIER ALPHA JET

Countries of Origin: France and Federal Germany.
Type: Tandem two-seat basic/advanced trainer and light tactical support aircraft.
Power Plant: Two 2,975 lb st (1 350 kgp) SNECMA-Turboméca Larzac 04-C5 or (NGEA) 3,360 lb st (1 525 kgp) 04-C20 turbofans.
Performance: (Larzac 04-C5) Max. speed (clean aircraft), 622 mph (1 000 km/h) or Mach 0·82 at sea level, 567 mph (912 km/h) or Mach 0·84 at 32,810 ft (10 000 m); max. climb, 11,220 ft/min (57 m/sec); tactical radius (training mission), 267 mls (430 km) at low altitude, 683 mls (1 100 km) at high altitude; ferry range (max. external fuel), 1,785 mls (2 872 km).
Weights: Empty, 7,716 lb (3 500 kg); loaded (clean), 11,023 lb (5 000 kg); max. take-off, 15,983 lb (7 250 kg).
Armament: (Close air support) External centreline gun pod with (Alpha Jet E) 30-mm DEFA 533 or (Alpha Jet A) 27-mm Mauser cannon, plus up to 4,850 lb (2 200 kg) of ordnance on four wing stations.
Status: First prototype flown 26 October 1973, with 410 delivered by beginning of 1983 against 487 ordered as follows: France (175), Germany (175), Belgium (33), Cameroun (6), Egypt (45), Ivory Coast (6), Morocco (24), Nigeria (12), Qatar (6) and Togo (5). Final assembly lines in Toulouse, Munich and (for Egyptian Air Force) Helwan.
Notes: Manufactured under joint Franco-German programme with French version optimised for training and German version for close air support. Alpha Jet NGEA (*Nouvelle Génération Ecole-Appui*) with enhanced weapon system entered flight test in April 1982. This version has been ordered by Egypt (15) and Cameroun (6). It features uprated Larzac engines and an integrated nav/attack system.

DASSAULT-BREGUET/DORNIER ALPHA JET

Dimensions: Span, 29 ft 11 in (9,11 m); length, 40 ft 3 in (12,29 m); height, 13 ft 9 in (4,19 m); wing area, 188 sq ft (17,50 m²).

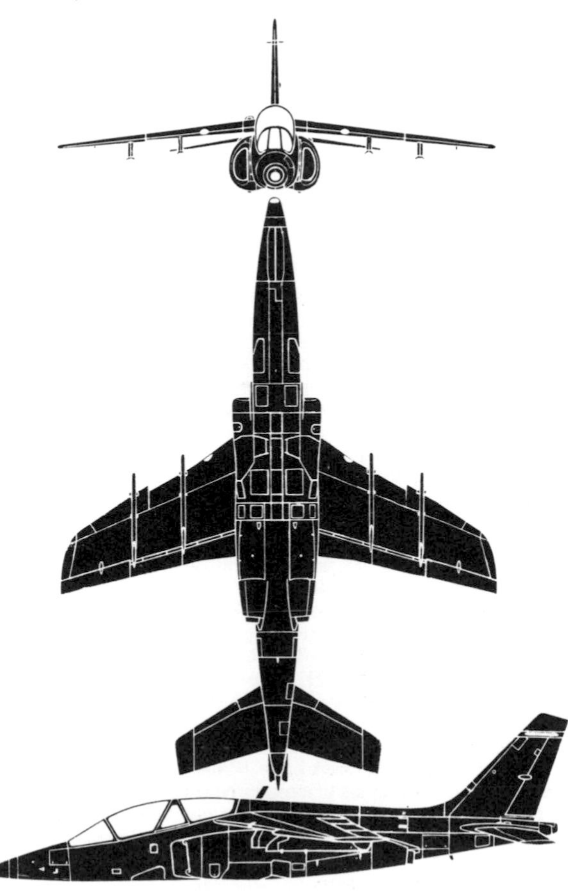

DE HAVILLAND CANADA DASH 7

Country of Origin: Canada.
Type: STOL regional airliner.
Power Plant: Four 1,120 shp Pratt & Whitney (Canada) PT6A-50 turboprops.
Performance: Max. cruising speed (with 9,500-lb/4 309-kg payload of 50 passengers and baggage), 266 mph (428 km/h) at 8,000 ft (2 440 m); normal cruise, 262 mph (421 km/h) at 15,000 ft (4 570 m); range cruise, 248 mph (399 km/h); range (50 passengers and IFR reserves), 795 mls (1 280 km) at normal cruise, (with max. fuel and 6,500-lb/2 948-kg payload), 1,347 mls (2 168 km).
Weights: Empty equipped, 27,600 lb (12 519 kg); max. take-off, 44,000 lb (19 958 kg).
Accommodation: Flight crew of two and standard seating for 50 passengers four-abreast with optional arrangement for 54 passengers or cargo/passenger mix.
Status: First of two pre-series aircraft flown 27 March 1975, with first production aircraft following on 30 May 1977. Some 90 delivered by beginning of 1983 against orders for approximately 130 and production running at one monthly.
Notes: Current basic aircraft is the Series 100, the all-cargo version being the Series 101, proposed versions including the Series 200 with 1,230 shp PT6A-55 engines and higher gross weights, and the Series 300 with an 18 ft 6 in (5,60 m) fuselage stretch to give a 50 per cent increase in passenger capacity. The last-mentioned version will have PT6A derivative engines of 1,500 shp. Two examples of a maritime surveillance version, the Dash 7R, have been supplied to the Canadian Coast Guard, and military operators of the Dash 7 are the Canadian Armed Forces (2) and the Venezuelan Navy (1). In Canadian military service the Dash 7 is designated CC-132.

DE HAVILLAND CANADA DASH 7

Dimensions: Span, 93 ft 0 in (28,35 m); length, 80 ft $7\frac{3}{4}$ in (24,58 m); height, 26 ft 2 in (7,98 m); wing area, 860 sq ft (79,90 m²).

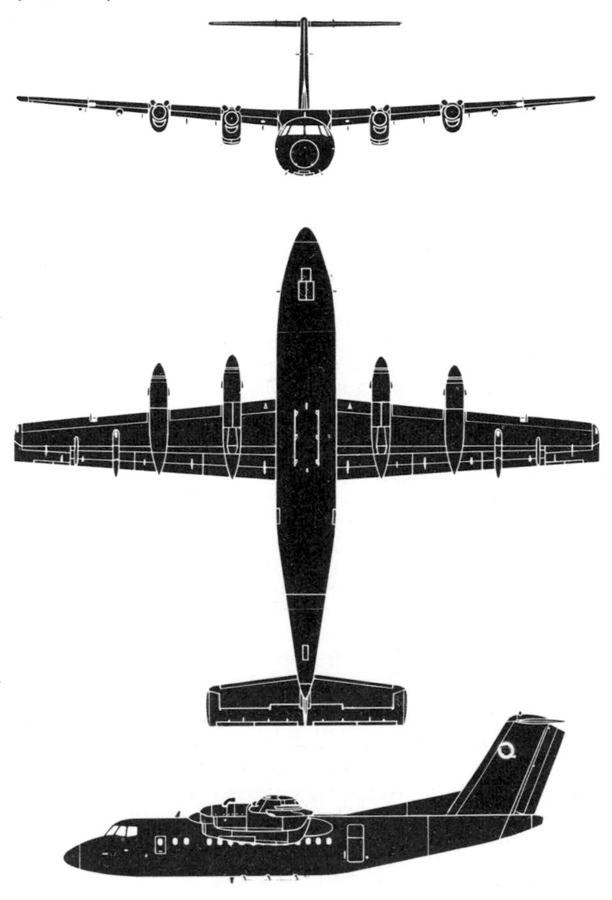

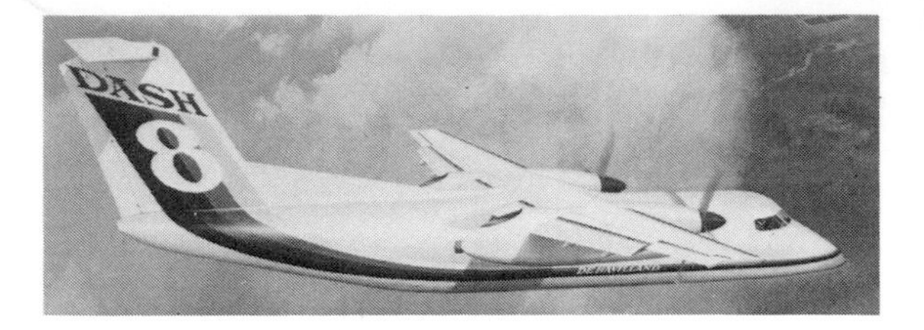

DE HAVILLAND CANADA DASH 8

Country of Origin: Canada.
Type: Regional airliner and corporate transport.
Power Plant: Two 1,800 shp Pratt & Whitney (Canada) PW120 turboprops.
Performance: Max. cruising speed, 311 mph (500 km/h) at 15,000 ft (4 570 m), 301 mph (484 km/h) at 25,000 ft (7 620 m); range (36 passengers at max. cruise with normal reserves), 380 mls (611 km) at 25,000 ft (7 620 m), (with 4,400-lb/1 996-kg payload at long-range cruise), 1,520 mls (2 446 km).
Weights: Operational empty, 20,176 lb (9 152 kg); max. take-off, 30,500 lb (13 835 kg).
Accommodation: Flight crew of two and standard arrangement for 36 passengers four-abreast with central aisle. Various alternative arrangements proposed including all first-class accommodation for 24 passengers and corporate versions with extended range capability and seating for about 17 passengers.
Status: First of four aircraft to participate in certification process scheduled to fly June 1983. Certification and first customer deliveries (to norOntair) planned for autumn 1984. Firm orders for 40 aircraft placed by 12 operators at beginning of 1983, with a further 78 on option.
Notes: The Dash 8 (DHC-8) is an evolutionary design embodying service-proven features of the four-engined Dash 7. The corporate transport version will incorporate an auxiliary power unit as standard and its extended range capability will enable 2,300 miles (3 700 km) to be flown with a 1,200-lb (544-kg) payload, a more typical mission being the transportation of 17 passengers and their baggage over a distance of 1,520 mls (2 446 km).

DE HAVILLAND CANADA DASH 8

Dimensions: Span, 84 ft 0 in (25,60 m); length, 73 ft 0 in (22,25 m); height, 25 ft 0 in (7,62 m); wing area, 585 sq ft (54,35 m²).

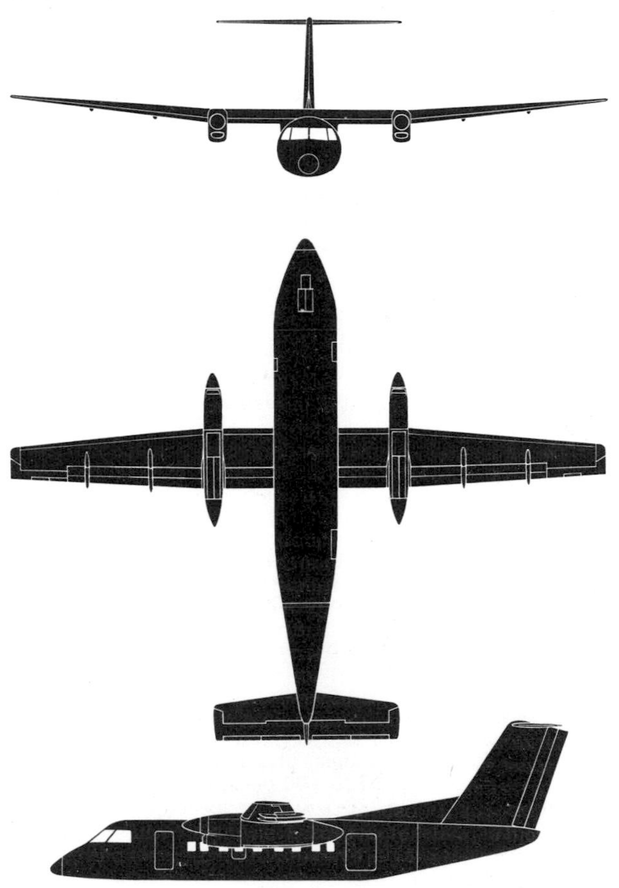

DORNIER DO 228

Country of Origin: Federal Germany.
Type: Light regional airliner and utility transport.
Power Plant: Two 715 shp Garrett AiResearch TPE 331-5 turboprops.
Performance: Max. cruising speed, 268 mph (432 km/h) at 10,000 ft (3 280 m), 230 mph (370 km/h) at sea level; initial climb, 2,050 ft/min (10,4 m/sec); service ceiling, 29,600 ft (9 020 m); range (-100), 1,224 mls (1 970 km) at max. range cruise, 1,075 mls (1 730 km) at max. cruise, (-200), 715 mls (1 150 km) at max. range cruise, 640 mls (1 030 km) at max. cruise.
Weights: Operational empty (-100), 7,132 lb (3 235 kg), (-200), 7,450 lb (3 379 kg); max. take-off, 12,570 lb (5 700 kg).
Accommodation: Flight crew of two and standard arrangements for (-100) 15 and (-200) 19 passengers in individual seats with central aisle.
Status: Prototype Do 228-100 flown on 28 March and -200 on 9 May 1981, and first customer delivery (A/S Norving) August 1982, with four more (-100s) delivered by beginning of 1983. Some 30 Do 228s (both -100s and -200s) ordered by beginning of 1983, in which year production is scheduled to reach two–three aircraft monthly.
Notes: The Do 228 mates a new-technology wing of supercritical section with the fuselage cross-section of the Do 128 (see 1982 edition), and two versions differing essentially in fuselage length and range capability are currently in production, the shorter-fuselage Do 228-100 being illustrated above and the longer-fuselage Do 228-200 being illustrated on the opposite page. All-cargo and corporate transport versions of the -100 are being offered, and geosurvey, maritime surveillance and military transport versions are proposed.

DORNIER DO 228

Dimensions: Span, 55 ft 7 in (16,97 m); length (-100) 49 ft 3 in (15,03 m), (-200), 54 ft 3 in (16,55 m); height, 15 ft 9 in (4,86 m); wing area, 344·46 sq ft (32,00 m²).

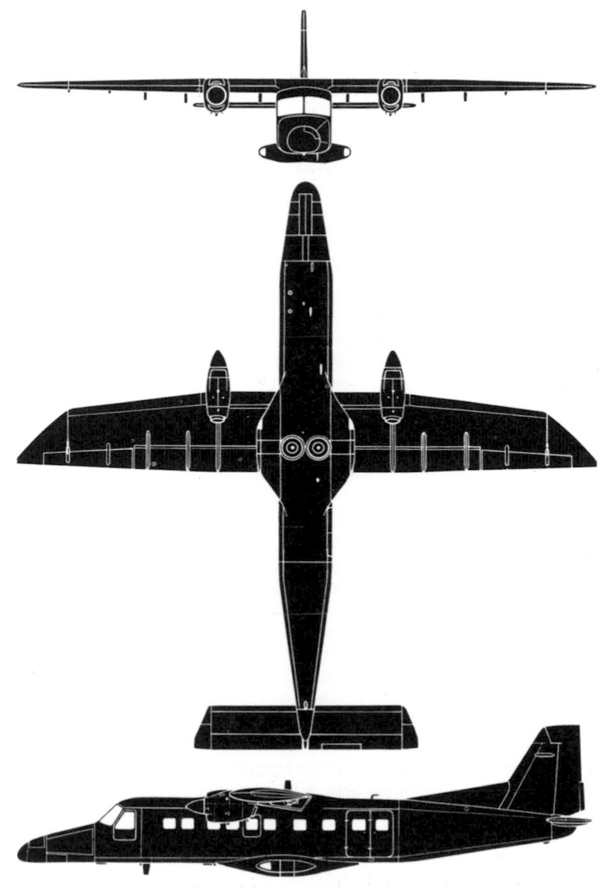

EDGLEY EA7 OPTICA

Country of Origin: United Kingdom.
Type: Three-seat observation aircraft.
Power Plant: One 200 hp Avco Lycoming IO-360 or 210 hp TIO-360 four-cylinder horizontally-opposed engine driving a ducted fan.
Performance: Max. speed, 126 mph (203 km/h); cruise (65% power), 108 mph (174 km/h); loiter speed, 57 mph (92 km/h); initial climb, 720 ft/min (3,66 m/sec); service ceiling, 14,000 ft (4 270 m); range (65% power), 737 mls (1 186 km); max. endurance at loiter speed, 10 hrs.
Weights: Empty, 1,875 lb (850 kg); max. take-off, 2,725 lb (1 236 kg).
Status: Prototype flown on 14 December 1979. Initial order for 25 placed mid-1981, and preparations initiated for production of 70 Opticas annually commencing mid-1983. Orders for more than 50 Opticas were at advanced stages of discussion at the beginning of 1983.
Notes: Of unique concept, the Optica is intended primarily for pipeline and powerline inspection, traffic surveillance, forestry, coastal and frontier patrol, and aerial photography, tasks that it is claimed capable of performing at less than one-third of the initial and operating costs of a comparable helicopter normally utilised for such. The engine is part of a ducted propulsor unit which forms a power pod separate from the main shroud and mounted downstream of a five-bladed fixed-pitch fan. The Optica can take-off and clear 50 ft (15 m) within 655 ft (200 m) and land from 50 ft within 850 ft (260 m).

EDGLEY EA7 OPTICA

Dimensions: Span, 39 ft 4 in (12,00 m); length, 26 ft $9\frac{1}{4}$ in (8,16 m); height, 6 ft $3\frac{1}{2}$ in (1,92 m); wing area, 170·5 sq ft (15,84 m²).

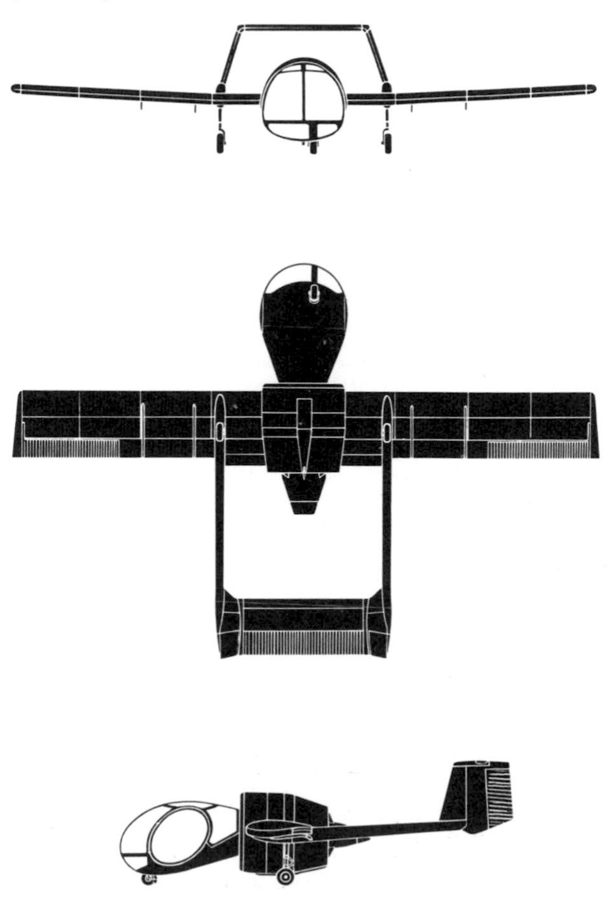

EMBRAER EMB-120 BRASILIA

Country of Origin: Brazil.
Type: Regional airliner, freighter and corporate transport.
Power Plant: Two 1,500 shp Pratt & Whitney PW115 turbo-props.
Performance: (Max. take-off weight) Max. cruising speed, 332 mph (534 km/h) at 20,000 ft (6 095 m); long-range cruise, 290 mph (467 km/h) at 20,000 ft (6 095 m); initial climb, 2,620 ft/min (13,30 m/sec); service ceiling, 32,000 ft (9 755 m); range (3C passengers and normal reserves), 662 mls (1 065 m); max. range (16 passengers), 1,808 mls (2 910 km).
Weights: Operational empty, 12,293 lb (5 576 kg); max. take-off, 21,164 lb (9 600 kg).
Accommodation: Flight crew of two (with provision for third member) and various arrangements for 24, 26 and 30 passengers three-abreast with offset aisle. All-freight version (7,006-lb/3 178-kg payload) and corporate transport versions (12 passengers in standard arrangement) proposed.
Status: First prototype scheduled to enter flight test on 29 July 1983, with certification expected in the second half of 1984, and first customer deliveries from mid-1985. Orders and options for approximately 120 aircraft from 23 operators in eight countries (i.e., Australia, Brazil, Colombia, Finland, France, Mexico, United Kingdom and USA) claimed by beginning of 1983.
Notes: Several military variants of the Brasilia were projected at the beginning of 1983, these including versions for maritime surveillance, electronic intelligence, aeromedical evacuation, paratroop transport, and search and rescue missions.

EMBRAER EMB-120 BRASILIA

Dimensions: Span, 64 ft 10¾ in (19,78 m); length, 65 ft 7½ in (20,00 m); height, 20 ft 10 in (6,35 m); wing area, 409·36 sq ft (38,03 m²).

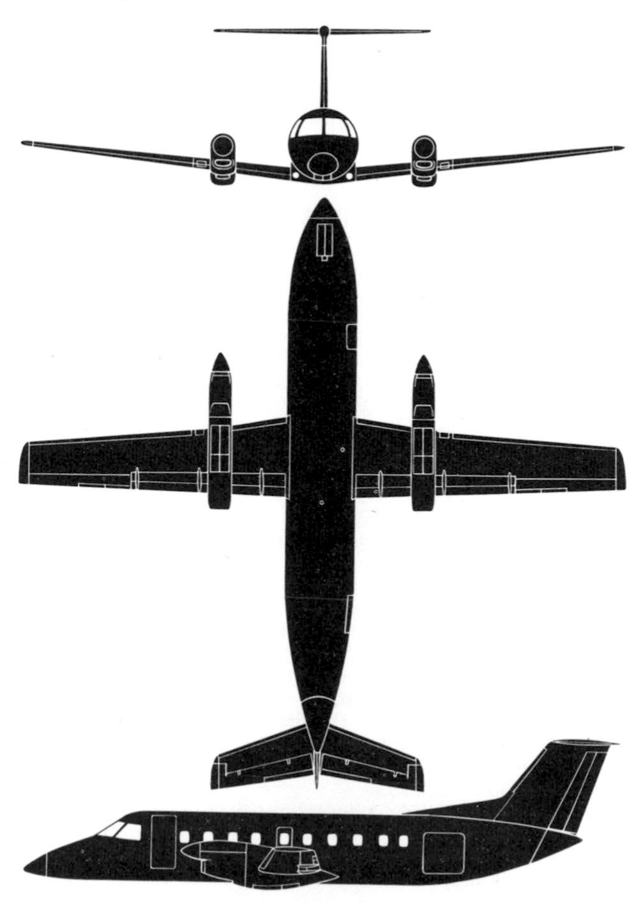

EMBRAER EMB-121A1 XINGU II

Country of Origin: Brazil.
Type: Light corporate transport and crew trainer.
Power Plant: Two 750 shp Pratt & Whitney (Canada) PT6A-135 turboprops.
Performance: (At 11,023 lb/5 000 kg) Max. cruising speed, 298 mph (480 km/h) at 12,000 ft (3 660 m), 289 mph (465 km/h) at 20,000 ft (6 100 m); initial climb, 1,820 ft/min (9,24 m/sec); service ceiling, 30,000 ft (9 145 m); range (max. payload at long-range cruise with 45 min reserves), 1,490 mls (2 400 km) at 30,000 ft (9 145 m).
Weights: Basic empty, 8,200 lb (3 720 kg); max. take-off, 12,500 lb (5 670 kg).
Accommodation: Two seats side-by-side on flight deck with alternative arrangements for five or six passengers in individual seats in main cabin.
Status: Prototype Xingu flown on 10 October 1976, with first production aircraft following on 20 May 1977, and 56 delivered by beginning of 1983, when sales totalled 99 aircraft and production was being increased from 1·5 to 2·5 aircraft per month to meet orders for 25 from the French *Armée de l'Air* and 16 for the French *Aéronavale* placed 16 January 1981 with deliveries commencing March 1982. The first Xingu II was flown on 4 September 1981, and this had replaced the initial model (apart from those ordered by the French services) on the assembly line by the beginning of 1983.
Notes: The Xingu II is a progressive development of the initial production version of the Xingu, with more powerful engines driving four-bladed propellers. Development of the stretched EMB-121B Xingu III was continuing at the beginning of 1983.

EMBRAER EMB-121A1 XINGU II

Dimensions: Span, 47 ft 5 in (14,45 m); length, 40 ft $2\frac{1}{4}$ in (12,25 m); height, 15 ft $6\frac{1}{2}$ in (4,74 m); wing area, 296 sq ft (27,50 m²).

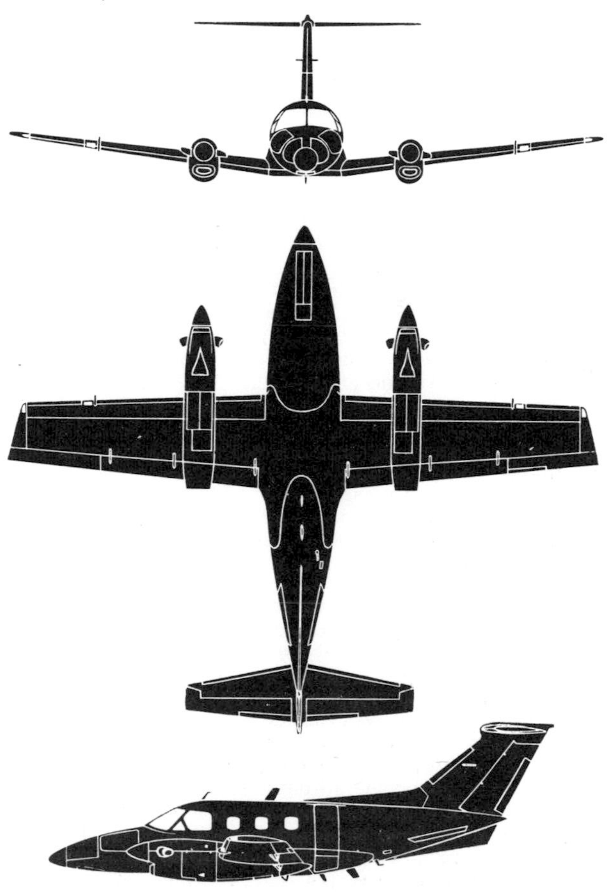

EMBRAER EMB-312 (T-27) TUCANO

Country of Origin: Brazil.
Type: Tandem two-seat basic and weapons trainer.
Power Plant: One 750 shp Pratt & Whitney (Canada) PT6A-25C turboprop.
Performance: (At 5,622 lb/2 550 kg) Max. speed, 291 mph (469 km/h) at 9,000 ft (2 745 m); econ cruise, 214 mph (345 km/h) at 10,000 ft (3 050 m); initial climb, 2,590 ft/min (13,15 m/sec); service ceiling, 28,500 ft (8 685 m); max. range (internal fuel), 1,179 mls (1 897 km).
Weights: Empty equipped, 3,487 lb (1 582 kg); max. take-off (aerobatic), 5,622 lb (2 550 kg), (with external stores), 7,000 lb (3 175 kg).
Armament: (Weapons training and light strike) Four under-wing stations for total weapons load of 1,323 lb (600 kg) which may comprise two 0·5-in (12,7-mm) gun pods, four pods each with seven 37-mm or 70-mm rockets, or four 250-lb (113-kg) bombs.
Status: First of four prototypes flown on 16 August 1980, with deliveries against Brazilian Air Force order for 118 aircraft (plus option on further 50) scheduled to commence early 1983, with production rate of five monthly by mid-year increasing to eight monthly by 1984.
Notes: The Tucano (Toucan) is unique among current production turboprop trainers in having been designed for this type of power plant from the outset and in having ejection seats. It simulates the flying characteristics of a pure jet aircraft which can thus be omitted from the basic wings syllabus of the Brazilian Air Force. An option on three Tucanos has been taken by CSE in the United Kingdom. The Tucano will initially replace the Cessna T-37 pure jet trainer in service with the Brazilian Air Force.

EMBRAER EMB-312 (T-27) TUCANO

Dimensions: Span, 36 ft $6\frac{1}{2}$ in (11,14 m); length, 32 ft $4\frac{1}{4}$ in (9,86 m); height, 11 ft $1\frac{7}{8}$ in (3, 40 m); wing area, 208·82 sq ft (19,40 m²).

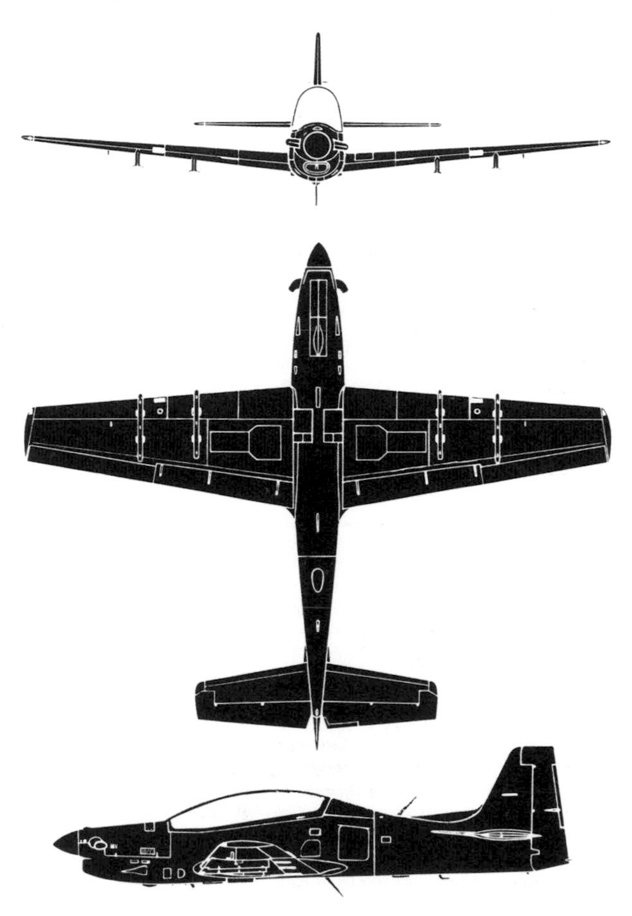

FOKKER F28 FELLOWSHIP MK 4000

Country of Origin: Netherlands.
Type: Short/medium-haul commercial airliner.
Power Plant: Two 9,850 lb st (4 468 kgp) Rolls-Royce RB. 183-2 Spey Mk 555-15H turbofans.
Performance: Max. cruising speed, 523 mph (843 km/h) at 23,000 ft (7 000 m); econ cruise, 487 mph (783 km/h) at 32,000 ft (9 755 m); range cruise, 421 mph (678 km/h) at 30,000 ft (9 145 m); range (with max. payload), 1,160 mls (1 870 km), (with max. fuel), 2,566 mls (4 130 km); cruise altitude, 35,000 ft (10 675 m).
Weights: Operational empty, 38,825 lb (17 661 kg); max. take-off, 73,000 lb (33 110 kg).
Accommodation: Flight crew of two (with jump seat for optional third crew member) and basic main cabin single-class configuration for 85 passengers five-abreast.
Status: First of two F28 prototypes flown on 8 May 1967, with first customer delivery following on 24 February 1969. Total of 189 ordered by beginning of 1983, with 177 delivered and production running at 1·0 to 1·5 monthly.
Notes: The F28 Mk 4000, which first flew in October 1976, provides the bulk of current production, the Mks 3000 and 4000 having supplanted the Mks 1000 and 2000 (after completion of 97 and 10 respectively). Whereas the Mk 3000 has the 80 ft 6½ in (24,55 m) fuselage of the Mk 1000, offering seating for up to 65 passengers, the Mk 4000 has the lengthened fuselage introduced by the Mk 2000. Both versions have the extended wing originally developed for the (subsequently discarded) Mks 5000 and 6000. Consideration is being given to the possible re-engining of the F28.

FOKKER F28 FELLOWSHIP MK 4000

Dimensions: Span, 82 ft 3 in (25,07 m); length, 97 ft $1\frac{3}{4}$ in (29,61 m); height, 27 ft $9\frac{1}{2}$ in (8,47 m); wing area, 850 sq ft (78,97 m²).

GATES LEARJET MODEL 55

Country of Origin: USA.
Type: Light corporate executive transport.
Power Plant: Two 3,700 lb st (1 678 kgp) Garrett AiResearch TFE 731-3A-2B turbofans.
Performance: Max. cruising speed, 525 mph (845 km/h) at 41,000 ft (12 495 m); time to 41,000 ft (12 495 m), 24 min; max. altitude, 51,000 ft (15 545 m); range (four passengers and 45 min reserves), 2,666 mls (4 290 km), (with 2,000 lb/907 kg payload), 2,073 mls (3 335 km).
Weights: Typical empty equipped, 12,600 lb (5 715 kg); max. take-off, 19,500 lb (8 845 kg), (optional), 20,500 lb (9 299 kg).
Accommodation: Flight crew of two and various main cabin arrangements for up to a maximum of 11 passengers. Typical arrangements include five individual seats and a bench-type seat for two, and four individual seats and a bench-type seat for four.
Status: The first of two prototypes flew on 19 April 1979, with first production aircraft following on 11 August 1980. Customer deliveries commenced 30 April 1981, some 65 having been delivered by beginning of 1983, production of four aircraft monthly being planned for that year.
Notes: Originally known as the Longhorn, the Model 55 is available in extended-range 55ER, 55LR and 55XLR versions, fuel tankage replacing the tail baggage compartment and, in the case of the special-mission 55XLR, an aft cabin tank being provided to boost range to 3,340 mls (5 375 km).

GATES LEARJET MODEL 55

Dimensions: Span, 43 ft 9½ in (13,34 m); length, 55 ft 1½ in (16,79 m); height, 14 ft 8 in (4,47 m); wing area, 264·5 sq ft (24,57 m²).

GENERAL DYNAMICS F-16 FIGHTING FALCON

Country of Origin: USA.
Type: (F-16A) Single-seat multi-role fighter and (F-16B) two-seat operational trainer.
Power Plant: One 14,800 lb st (6 713 kgp) dry and 23,830 lb st (10 809 kgp) reheat Pratt & Whitney F100-PW-200 turbofan.
Performance: Max. speed (short endurance dash), 1,333 mph (2 145 km/h) or Mach 2·02, (sustained), 1,247 mph (2 007 km/h) or Mach 1·89 at 40,000 ft (12 190 m); max. cruise, 614 mph (988 km/h) or Mach 0·93; tactical radius (HI-LO-HI interdiction on internal fuel), 360 mls (580 km) with six 500-lb (227-kg) bombs; range (similar ordnance load and internal fuel), 1,200 mls (1 930 km) at 575 mph (925 km/h).
Weights: Operational empty, 14,567 lb (6 613 kg); max. take-off, 35,400 lb (16 057 kg).
Armament: One 20-mm M61A-1 multi-barrel rotary cannon and from two to six AIM-9L/M AAMs, or (air support) up to 12,000 lb (5 443 kg) of ordnance between nine stations.
Status: First of two (YF-16) prototypes flown 20 January 1974. First production F-16 flown 7 August 1978, with some 860 delivered by 1983 by parent company (10 monthly) and European consortium (five monthly), the latter having final assembly lines in Belgium and Netherlands. USAF procurement calls for 785 F-16A/Bs and 704 F-16C/Ds. European programme embraces 116 for Belgium, 58 for Denmark, 142 for Netherlands and 72 for Norway. Export orders comprise Israel (75), Egypt (40), Pakistan (40) and South Korea (36).
Notes: Single-seat F-16C and two-seat F-16D with upgraded systems to be delivered from July 1984. The F-16/J79 for export with the J79 engine and the F-16/101 with F-101DFE engine flown 29 October and 19 December 1980.

GENERAL DYNAMICS F-16 FIGHTING FALCON

Dimensions: Span (excluding missiles), 31 ft 0 in (9,45 m); length, 47 ft 7¾ in (14,52 m); height, 16 ft 5¼ in (5,01 m); wing area, 300 sq ft (27,87 m²).

GENERAL DYNAMICS F-16XL

Country of Origin: USA.
Type: Advanced multi-role fighter technology demonstrator.
Power Plant: (1st F-16XL) One 14,800 lb st (6 713 kgp) dry and 23,830 lb st (10 809 kgp) reheat Pratt & Whitney F100-PW-200, or (2nd F-16XL) one 29,000 lb st (13 154 kgp) reheat General Electric F-110 turbofan.
Performance: (Estimated) Max. speed (short endurance dash), 1,650 mph (2 655 km/h) or Mach 2·5 at 40,000 ft (12 190 m); max. cruise, 1,452 mph (2 337 km/h) or Mach 2·2 above 40,000 ft (12 190 m); tactical radius (HI-LO-HI interdiction on internal fuel with 8,000 lb/3 629 kg ordnance), 520 mls (837 km), (with 4,000 lb/1 814 kg ordnance), 805 mls (1 295 km); max. range, 2,880 mls (4 635 km) plus.
Weights: Design mission, 43,000 lb (19 505 kg); max. take-off, 48,000 lb (21 773 kg).
Armament: (Air Combat) Ten AMRAAM (Advanced Medium-Range Air-to-Air Missile), or (air support) up to 15,000 lb (6 804 kg) of bombs and missiles.
Status: First (single-seat) F-16XL flown 3 July 1982, and second (two-seat) on 29 October 1982. Both prototypes to undergo USAF Aeronautical Systems Division evaluation after completion of manufacturer's test programme in April 1983, with view to series production (as the F-16E).
Notes: Featuring a wing of cranked arrow configuration, the F-16XL is a derivative of the F-16 Fighting Falcon (see page 94). The wings are of graphite composite and increase internal fuel capacity by 82 per cent, and the fuselage, while essentially similar to that of the standard F-16, embodies two plugs totalling 56 in (1,42 m). The second (two-seat) F-16XL is illustrated above.

GENERAL DYNAMICS F-16XL

Dimensions: Span, 34 ft 2$\frac{4}{5}$ in (10,43 m); length, 54 ft 1$\frac{7}{8}$ in (16,51 m); height, 17 ft 7 in (5,36 m).

GRUMMAN E-2C HAWKEYE

Country of Origin: USA.
Type: Airborne early warning, surface surveillance and strike control aircraft.
Power Plant: Two 4,910 ehp Allison T56-A-425 turboprops.
Performance: Max. speed, 348 mph (560 km/h) at 10,000 ft (3 050 m); max. range cruise, 309 mph (498 km/h); initial climb, 2,515 ft/min (12,8 m/sec); service ceiling, 30,800 ft (9 390 m); mission endurance (at 230 mls/370 km from base), 4·0 hrs; max. endurance, 6·1 hrs; ferry range, 1,604 mls (2 580 km).
Weights: Empty, 38,009 lb (17 240 kg); max. take-off, 51,900 lb (23 540 kg).
Accommodation: Crew of five comprising flight crew of two and Airborne Tactical Data System team of three, each occupying an independent operating station.
Status: First of two E-2C prototypes flown on 20 January 1971, with first production aircraft flying on 23 September 1972. Total US Navy requirement for 101 by 1986, with 85 funded through Fiscal Year 1982. Four delivered to Israel (one of which is illustrated above), and first two of eight ordered by Japan to reach that country in February 1983. First of four E-2Cs for Egypt expected to be delivered during 1985.
Notes: The E-2C is the current production version of the Hawkeye, having followed 59 E-2As (all subsequently updated to E-2B standards), and is able to operate independently, in co-operation with other aircraft, or in concert with a ground environment. Two have been delivered to the US Navy as TE-2Cs for use as conversion trainers by the Service's two Hawkeye readiness squadrons which support 12 four-aircraft Hawkeye squadrons attached to the carrier air wings. Current planning calls for E-2C production to continue through 1986.

GRUMMAN E-2C HAWKEYE

Dimensions: Span, 80 ft 7 in (24,56 m); length, 57 ft 7 in (17,55 m); height, 18 ft 4 in (5,69 m); wing area, 700 sq ft (65,03 m²).

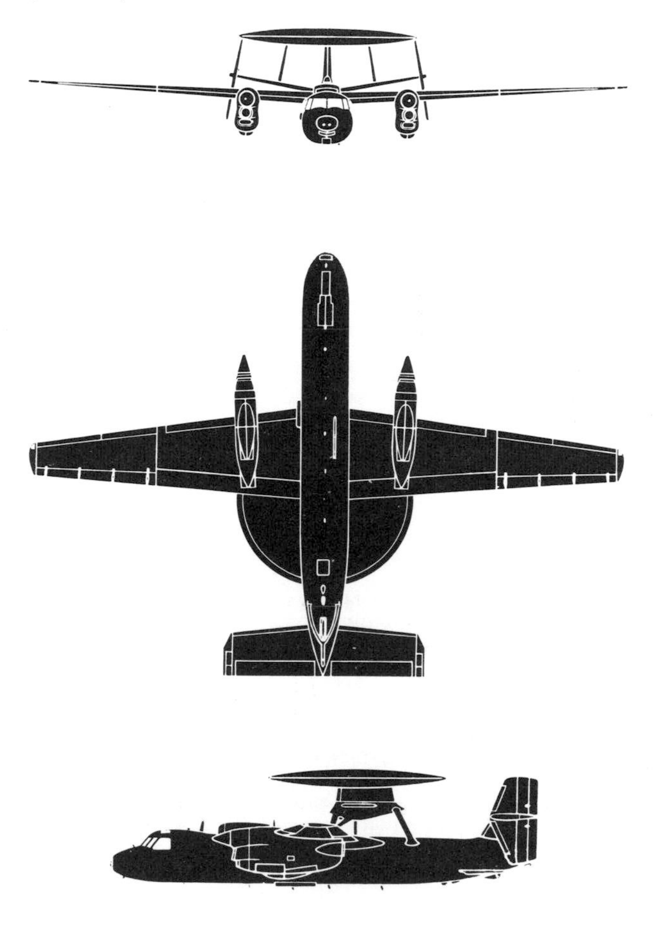

GRUMMAN F-14A TOMCAT

Country of Origin: USA.
Type: Two-seat shipboard multi-role fighter.
Power Plant: Two 12,500 lb st (5 670 kgp) dry and 20,900 lb st (9 840 kgp) reheat Pratt & Whitney TF30-P-412A or P-414 turbofans.
Performance: Max. speed (with four semi-recessed AIM-7 AAMs), 913 mph (1 470 km/h) or Mach 1·2 at sea level, 1,584 mph (2 549 km/h) or Mach 2·4 at 49,000 ft (14 935 m); time to 60,000 ft (18 290 m) at 55,000 lb (24 948 kg), 2·1 min; tactical radius (combat air patrol on internal fuel), 765 mls (1 232 km).
Weights: Empty, 39,930 lb (18 112 kg); loaded (intercept mission with four AIM-7s), 58,904 lb (26 718 kg), (with six AIM-54s), 69,790 lb (31 656 kg); max. take-off, 74,348 lb (33 724 kg).
Armament: One 20-mm M61A-1 rotary cannon and (intercept) six AIM-7E/F Sparrow and four AIM-9G/H Sidewinder AAMs, or six AIM-54A Phoenix and two AIM-9G/H AAMs.
Status: First of 12 research and development aircraft flown 21 December 1970, with more than 400 delivered to US Navy by beginning of 1983, when production was expected to continue at a rate of 30 annually through Fiscal 1994 for an eventual total of 845 aircraft. Forty-nine being delivered as RF-14As for photo-recce role.
Notes: An engine upgrading programme (commencing with new aircraft), together with avionics and radar improvement programmes scheduled to commence 1985. Re-engined F-14s will have either General Electric F110 or Pratt & Whitney PW1130 turbofans.

GRUMMAN F-14A TOMCAT

Dimensions: Span (20 deg sweep), 64 ft 1½ in (19,55 m); (68 deg sweep), 37 ft 7 in (11,45 m); length, 61 ft 11⅞ in (18,90 m); height, 16 ft 0 in (4,88 m); wing area, 565 sq ft (52,50 m²).

GULFSTREAM AEROSPACE GULFSTREAM SMA-3

Country of Origin: USA.
Type: Light transport and multi-mission aircraft.
Power Plant: Two 11,400 lb st (5 171 kgp) Rolls-Royce RB.163–25 Spey Mk 511-8 turbofans.
Performance: Max. cruising speed, 577 mph (928 km/h) or Mach 0·85 at 30,000 ft (9 145 m); long-range cruise, 512 mph (825 km/h) or Mach 0·775; initial climb, 3,800 ft/min (19,3 m/sec); max. operating altitude, 45,000 ft (13 715 m); range (VFR reserves and 1,600 lb/726 kg payload), 4,720 mls (7 595 km), (IFR reserves), 4,200 mls (6 760 km).
Weights: Manufacturer's bare empty, 36,346 lb (16 486 kg); max. take-off, 68,200 lb (30 936 kg).
Accommodation: Flight crew of two or three and (personnel transport) up to 19 passengers or (medical evacuation) 15 litter patients and two medical attendants. For maritime surveillance a flight crew of three and three systems operators are carried, and in the cargo role up to 4,630 lb (2 100 kg) may be carried.
Status: Three Gulfstream SMA-3s were delivered to the Royal Danish Air Force during 1982, this service having an option on two further aircraft. The SMA-3 is a multi-mission version of the Gulfstream III corporate executive transport first flown on 2 December 1979. Customer deliveries commenced in September 1980, and 75 delivered by beginning of 1983, with 36 scheduled for delivery during the course of the year.
Notes: The Gulfstream SMA-3 differs from the basic Gulfstream III (see 1982 edition) primarily in having a large freight door in the starboard side and APS-127 surveillance radar. The examples supplied to the Royal Danish Air Force can be converted within two hours from high-density personnel transport configuration to VIP transport, cargo transport, maritime surveillance aircraft or search-and-rescue aircraft.

GULFSTREAM AEROSPACE GULFSTREAM SMA-3

Dimensions: Span, 77 ft 10 in (23,72 m); length 83 ft 1 in (25,30 m); height, 24 ft 4½ in (7,40 m); wing area, 934·6 sq ft (86,82 m²).

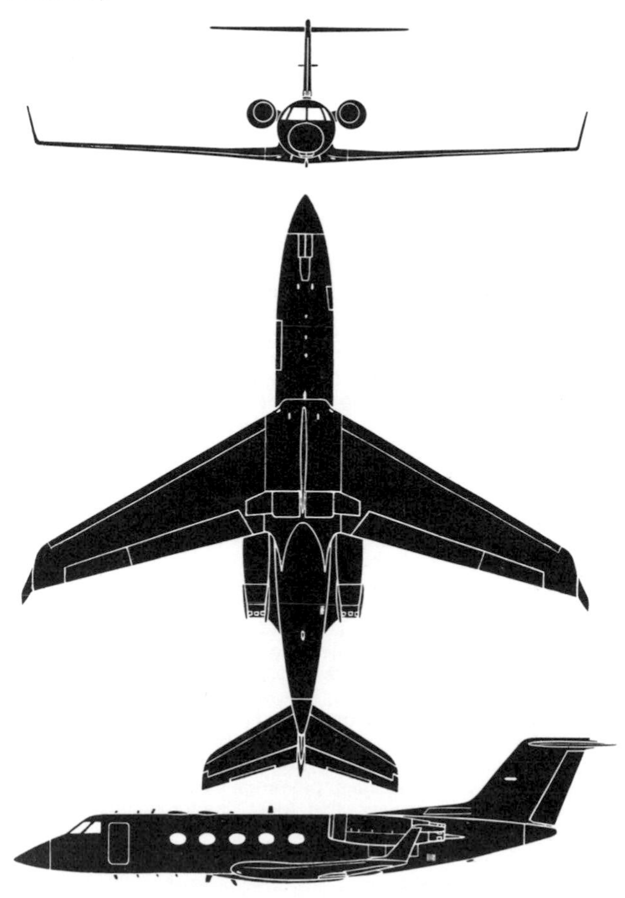

HAL AJEET TRAINER

Country of Origin: India.
Type: Tandem two-seat advanced trainer.
Power Plant: One 4,500 lb st (2 041 kgp) HAL-manufactured Rolls-Royce Orpheus 701-01 turbojet.
Performance: Max. speed, 665 mph (1 070 km/h) at sea level, 590 mph (948 km/h) at 36,090 ft (11 000 m); initial climb rate, 3,240 ft/min (16,46 m/sec); time to 39,370 ft (12 000 m), 7·5 min; ferry range (with reserves and two 29·7 Imp gal/135 l drop tanks), 560 mls (900 km) at average speed of 480 mph (772 km/h) at 39,370 ft (12 000 m).
Weights: Empty, 5,686 lb (2 579 kg); loaded (clean), 8,000 lb (3 629 kg); max. take-off, 9,683 lb (4 392 kg).
Armament: Two 30-mm Aden cannon and two 500-lb (226,8-kg) bombs, cluster bombs, 57-mm rocket pods or practice bomb packs each containing four 25-lb (11,34-kg) bombs on inboard stations (plus two 29·7 Imp gal/135 l drop tanks on outboard stations).
Status: Prototype Ajeet Trainer flown on 20 September 1982, and initial batch of 12 ordered by Indian Air Force as replacement for Hunter at OCU from 1984. Indian Navy has an initial requirement for eight Ajeet Trainers.
Notes: The Ajeet Trainer is a tandem two-seat training derivative of the single-seat Ajeet lightweight fighter (see 1978 edition), which, in turn, was an improved version of the licence-built Folland Gnat. The first Ajeet (Unconquerable) prototype, adapted from the 215th HAL-built Gnat, was flown on 6 March 1975, and 79 Ajeets built (plus 10 Gnats converted to similar standards) when production terminated in 1981. The Ajeet is now assigned the short-range close air support role in Indian Air Force service. The Ajeet Trainer retains the main hydraulic system and powered flying controls of the single-seater. It is claimed to possess a lower unit price than any other trainer in its category.

HAL AJEET TRAINER

Dimensions: Span, 22 ft 1 in (6,73 m); length, 34 ft $3\frac{1}{2}$ in (10,45 m); height, 8 ft $5\frac{1}{3}$ in (2,57 m); wing area, 136·12 sq ft (12,65 m²).

HAL HPT-32

Country of Origin: India.
Type: Side-by-side two-seat primary trainer.
Power Plant: One 260 hp Avco Lycoming AEIO-540-D4B5 six-cylinder horizontally-opposed engine.
Performance: Max. speed, 157 mph (253 km/h) at sea level; initial climb rate, 1,100 ft/min (5,6 m/sec); service ceiling, 15,580 ft (4 750 m); range (standard internal fuel), 490 mls (790 km) at 10,000 ft (3 050 m).
Weights: Empty, 1,940 lb (880 kg); max. take-off, 2,667 lb (1 210 kg).
Status: First of three prototypes flown on 6 January 1977, with definitive third prototype flying on 31 July 1981. Initial production series of 40 commenced 1982, with deliveries scheduled to commence during 1984 against total Indian Air Force requirement of up to 160 aircraft. Production rate of 20–30 annually anticipated from 1985.
Notes: The HPT-32 has been the subject of protracted development, the definitive form differing appreciably from the initial design as represented by the first prototype (see 1978 edition). During the course of development, the HPT-32 has undergone considerable structural revision to reduce weight, and the cockpit canopy, rear fuselage and vertical tail surfaces have been extensively redesigned. The HPT-32 is intended to provide grading and primary instruction after which pupils will convert to the HJT-16 Kiran. The basic design makes provision for the introduction of either one or two additional seats at the rear of the cabin, plus baggage space, to enable the HPT-32 to fulfil a secondary liaison and communications role, and a four-seat civil version is to be developed.

HAL HPT-32

Dimensions: Span, 31 ft 2 in (9,50 m); length, 25 ft 3¾ in (7,72 m); height, 9 ft 5½ in (2,88 m); wing area, 161·6 sq ft

HARBIN Y-11T TURBO-PANDA

Country of Origin: China.
Type: Light transport and general utility aircraft.
Power Plant: Two 475 shp Pratt & Whitney (Canada) PT6A-110 or (Y11T2) 620 shp PT6A-27 turboprops.
Performance: (Y-11T) Max. speed, 175 mph (282 km/h) at 10,000 ft (3 050 m); long-range cruise, 158 mph (254 km/h); operating speed (agricultural version), 100–112 mph (160–180 km/h); initial climb, 1,160 ft/min (5,89 m/sec); service ceiling, 22,965 ft (7 000 m); max. range (no reserves), 867 mls (1 410 km).
Weights: (Y-11T) Operational empty, 6,614 lb (3 000 kg); max. take-off, 12,125 lb (5 500 kg).
Accommodation: Flight crew of two and (passenger transport version), 15 passengers three-abreast. All-freight, mixed freight/passenger, agricultural, geophysical survey and other variants proposed.
Status: Following flight trials conducted in 1981 with turboprop-powered adaptation of Y-11, the first of two Y-11T production prototypes was flown on 14 July 1982. Initial deliveries of definitive version (Y-11T2) scheduled to commence early 1984.
Notes: Derived from the piston-engined Y-11 (see 1981 edition) by the State Aircraft Factories at Harbin (Pinkiang) and being offered for export by CATIC (China National Aero Technology Import and Export Corporation) as the Turbo-Panda, the Y-11T has been designed to meet western certification requirements. The series model will differ from the prototype primarily in omitting wing leading-edge slats and in having the higher-rated -27 turboprops. Western avionics, instrumentation and consumable items are being fitted.

HARBIN Y-11T TURBO-PANDA

Dimensions: Span, 56 ft 6 in (17,23 m); length, 48 ft 9 in (14,86 m); height, 17 ft 3½ in (5,28 m); wing area, 368·89 sq ft (34,27 m²).

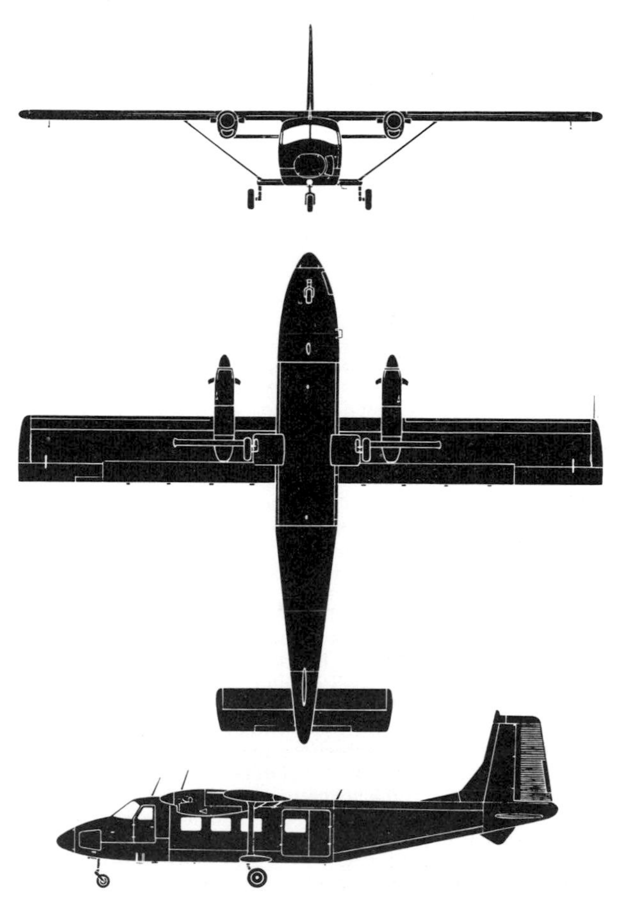

ICA IAR-825 TP TRIUMF

Country of Origin: Romania.
Type: Tandem two-seat basic/advanced trainer.
Power Plant: (Prototype) One 680 shp Pratt & Whitney (Canada) PT6A-15AG or (production) 550 shp PT6A-25 turboprop.
Performance: Max. speed, 292 mph (470 km/h); max. continuous cruising speed, 261 mph (420 km/h) at 6,560 ft (2 000 m); econ cruise, 242 mph (390 km/h) at 19,658 ft (6 000 m); initial climb, 2,362 ft/min (12 m/sec); service ceiling, 26,245 ft (8 000 m); endurance (internal fuel), 3·0 hrs.
Weights: Empty, 2,645 lb (1 200 kg); loaded (aerobatic), 3,307 lb (1 500 kg); normal loaded, 4,850 lb (2 200 kg).
Status: First prototype flown 12 June 1982, with series production scheduled to commence during first quarter of 1983, with an initial batch of 15 aircraft for evaluation by the Romanian Air Force which has a requirement for about 100 aircraft in this category.
Notes: The IAR-825 TP Triumf (Triumph) utilises a substantial proportion of the structure of the Lycoming IO-540-engined IAR-823 (see 1975 edition) which is currently used by the Romanian Air Force both as a side-by-side two-seat primary trainer and as a four/five-seat liaison aircraft. The first prototype is fitted with a PT6A-15AG engine previously flown in the IAR-827 TP turboprop-powered experimental version of the IAR-827 agricultural aircraft (see 1982 edition), and a second, definitive prototype was scheduled to enter flight test late 1982 or early 1983. The IAR-825 TP is expected to be adopted by the Romanian Air Force to carry pupil pilots from *ab initio* instruction on the IAR-823 through to the two-seat version of the Yugoslav-Romanian IAR-93 light close air support fighter (see page 196). The IAR-825 TP is also to be tested with the Czechoslovak M-601 B turboprop.

ICA IAR-825 TP TRIUMF

Dimensions: Span, 33 ft 9½ in (10,30 m); length 29 ft 2⅓ in (8,90 m); height, 7 ft 9¾ in (2,38 m).

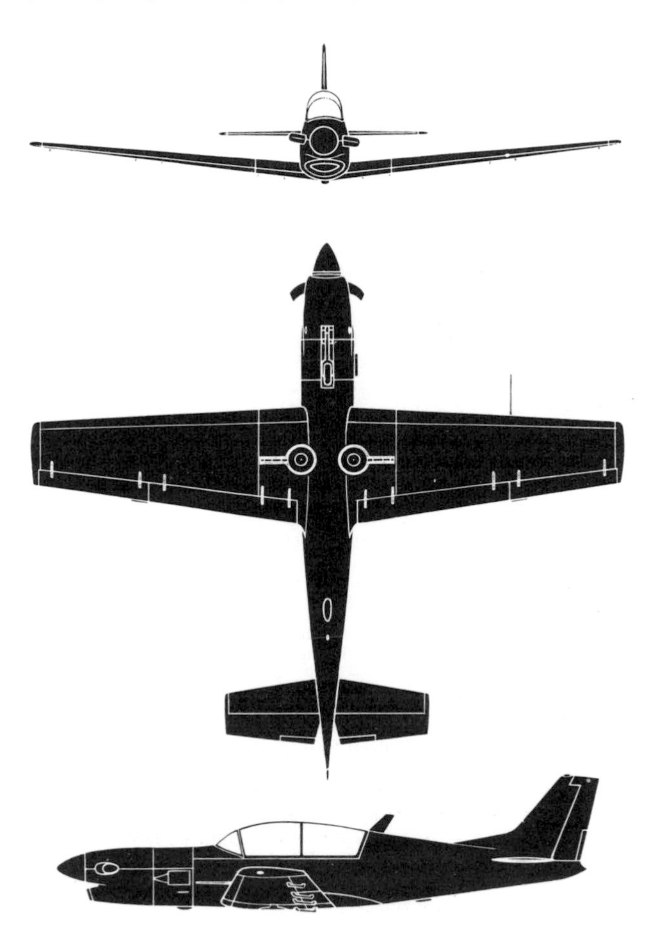

INDAER/FAC T-35 PILLAN

Country of Origin: Chile (USA).
Type: Tandem two-seat primary/basic trainer.
Power Plant: One 300 hp Avco Lycoming IO-540-K1 six-cylinder horizontally-opposed engine.
Performance: (At max. take-off weight) Max. speed, 193 mph (311 km/h) at sea level; cruise (75% power), 185 mph (298 km/h) at 8,800 ft (2 680 m), (65% power), 173 mph (278 km/h) at 12,800 ft (3 900 m); initial climb, 1,516 ft/min (7,7 m/sec); time to 6,000 ft (1 830 m), 4·7 min; range (with 45 min reserves), 680 mls (1 093 km) at 75% power, 720 mls (1 157 km) at 65% power.
Weights: Empty, 1,836 lb (832 kg); empty equipped, 2,048 lb (929 kg); max. take-off, 2,900 lb (1 315 kg).
Armament: (Light strike and armament training) Two pods of four or seven rockets, 250-lb (113,4-kg) bombs or 12,7-mm machine gun pods.
Status: First of two prototypes assembled by Piper at Lakeland flown spring 1981, with further three prototypes assembled by the FAC Maintenance Wing at El Bosque, Santiago. The FAC (*Fuerza Aérea de Chile*) has requirement for approximately 100 to be assembled by INDAER (not to be confused with Peruvian organisation of same name) with progressively increasing proportion of Chilean-manufactured parts. First production Pillan scheduled to fly early 1983, with 42 delivered by spring 1984.
Notes: Of joint FAC/Piper inception, the Pillan mates an entirely new tandem two-seat centre fuselage with the rear fuselage and reduced-span wings of the Piper PA-32R-301 Saratoga, the undercarriage of the PA-28R-200 Arrow and the flaps of the PA-32-300 Lance.

INDAER/FAC T-35 PILLAN

Dimensions: Span, 28 ft 11 in (8,81 m); length, 26 ft 1 in (7,97 m); height, 7 ft 8⅛ in (7,70 m); wing area, 147 sq ft (13,64 m²).

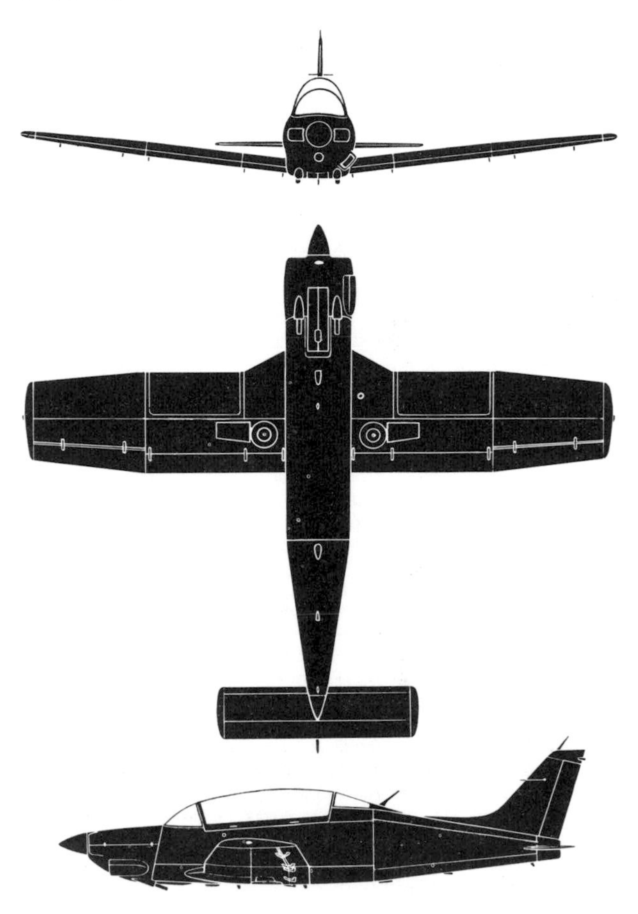

ILYUSHIN IL-76 (CANDID)

Country of Origin: USSR.
Type: Heavy-duty medium/long-haul military and commercial freighter and troop transport.
Power Plant: Four 26,455 lb st (12 000 kgp) Soloviev D-30KP turbofans.
Performance: Max speed, 528 mph (850 km/h) at 32,810 ft (10 000 m); max. cruise, 497 mph (800 km/h) at 29,500–42,650 ft (9 000–13 000 m); range cruise, 466 mph (750 km/h); range (with max. payload), 3,290 mls (5 300 km), (with max. fuel), 4,163 mls (6 700 km).
Weights: Max. take-off, 374,790 lb (170 000 kg).
Accommodation: Normal flight crew of four (with navigator below flight deck in glazed nose). Pressurised hold for up to 140 fully-equipped troops, containerised freight (up to 88,185 lb/40 000 kg), wheeled or tracked vehicles, self-propelled anti-aircraft guns, etc.
Armament: (Military version) Twin 23-mm cannon in tail barbette.
Status: First of four prototypes flown on 25 March 1971, with production deliveries to both Soviet Air Force and Aeroflot commencing in 1974. Soviet Air Force air transport component is believed to have received 150–160 Il-76s by beginning of 1983, when production was running at 35–40 annually.
Notes: The Il-76 was developed primarily for military roles but has had the distinction of being the first purpose-designed jet freighter to enter commercial service, operating with Aeroflot and in dual military/civil roles with Iraq, Libya and Syria. A flight-refuelling tanker version of the Il-76 is known to be entering Soviet Air Force service, and an airborne warning and control system version (AWACS) is expected to attain IOC (initial operational capability) 1984–85, this reportedly having a saucer-type rotating radome and a lengthened fuselage.

ILYUSHIN IL-76 (CANDID)

Dimensions: Span, 165 ft $8\frac{1}{3}$ in (50,50 m); length, 152 ft $10\frac{1}{4}$ in (46,59 m); height, 48 ft $5\frac{1}{8}$ in (14,76 m); wing area, 3,229·2 sq ft (300,00 m²).

ILYUSHIN IL-86 (CAMBER)

Country of Origin: USSR.
Type: Medium-haul commercial airliner.
Power Plant: Four 28,660lb st (13 000 kgp) Kuznetsov NK-86 turbofans.
Performance: Max. cruising speed, 590 mph (950 km/h) at 29,530 ft (9 000 m); econ cruise, 559 mph (900 km/h) at 36,090 ft (11 000 m); range (with max. payload—350 passengers), 2,485 mls (4 000 km), (with 250 passengers), 3,107 mls (5 000 km).
Weights: Max. take-off, 454,150 lb (206 000 kg).
Accommodation: Basic flight crew of three-four and up to 350 passengers nine-abreast with two aisles and divided between three cabins seating 111, 141 and 98 passengers.
Status: First prototype flown on 22 December 1976, and production prototype flown on 24 October 1977. Deliveries to Aeroflot commenced 1980, and some 25 are believed to have been delivered by the beginning of 1983. The Polish WSK-Mielec concern is responsible for manufacture of the entire wing, stabiliser and engine pylones. Four Il-86s are scheduled to be delivered to Polish Airlines LOT.
Notes: The Il-86 operated its first scheduled service (Moscow–Tashkent) on 26 December 1980, and first international service (Moscow–Prague) on 12 October 1981, but there has been an unexplained slippage in Aeroflot's programmed introduction of the Il-86 on many routes and it is believed that performance has fallen short of expectations. It is expected that a long-range version will be introduced once suitable engines are available.

ILYUSHIN IL-86 (CAMBER)

Dimensions: Span, 157 ft $8\frac{1}{8}$ in (48,06 m); length, 195 ft 4 in (59,54 m); height, 51 ft $10\frac{1}{2}$ in (15,81 m); wing area, 3,550 sq ft (329,80 m²).

LEAR FAN 2100

Country of Origin: USA.
Type: Light corporate executive transport.
Power Plant: Two 650 shp Pratt & Whitney (Canada) PT6B-35F turboshafts.
Performance: (Estimated) Max. cruising speed, 418 mph (673 km/h) at 20,000 ft (6 095 m), 358 mph (576 km/h) at 40,000 ft (12 190 m); long-range cruise, 267 mph (430 km/h) at 20,000 ft (6 095 m), 322 mph (518 km/h) at 40,000 ft (12 190 m); initial climb, 3,450 ft/min (17,53 m/sec); max. range (with 45 min reserves), 2,032 mls (3 270 km).
Weights: Empty, 4,100 lb (1 860 kg); max. take-off, 7,350 lb (3 334 kg).
Accommodation: Pilot and co-pilot/passenger on flight deck and various arrangements for up to eight passengers in main cabin.
Status: The first of two flying prototypes flew on 1 January 1981, and the second on 18 June 1982. Third and fourth aircraft for certification to fly April and May 1983 respectively. Certification planned for Summer 1983, with first customer deliveries following in September. First 40 production aircraft to be assembled at Reno, Nevada, subsequent aircraft being assembled at Aldergrove, Belfast, where 150 are scheduled to be produced in 1984 and 300 in 1985. Total of approximately 275 Lear Fan 2100s ordered by beginning of 1983.
Notes: The Lear Fan 2100 is manufactured primarily of composite materials, graphite/epoxy being used for the fuselage and all surfaces. The turboshafts drive the pusher propeller via a gearbox drive train and all fuel is housed in integral wing tanks.

LEAR FAN 2100

Dimensions: Span, 39 ft 4 in (11,99 m); length, 40 ft 7 in (12,37 m); height, 12 ft 2 in (3,70 m); wing area, 162·9 sq ft (15, 13 m²).

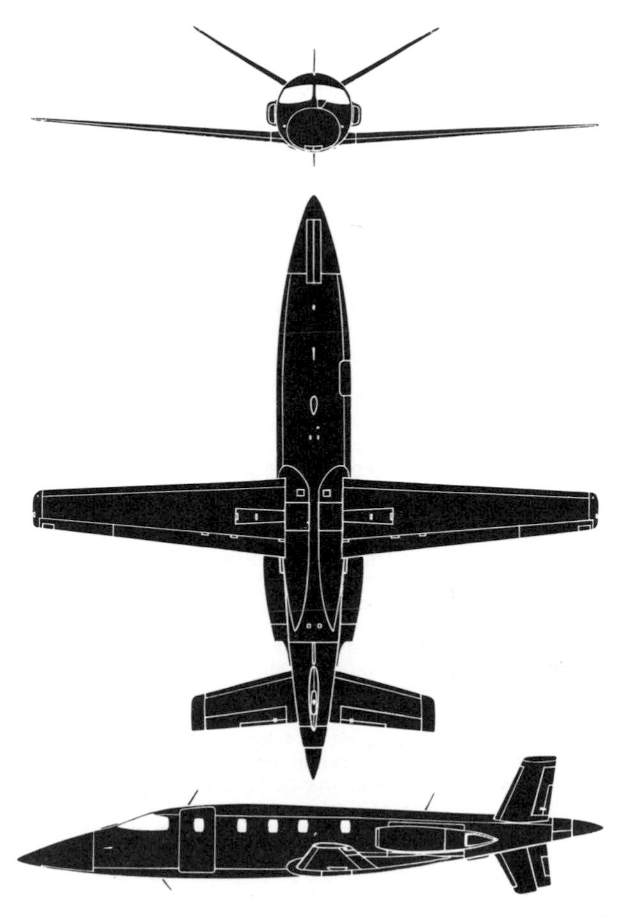

LOCKHEED L-100-30 HERCULES

Country of Origin: USA.
Type: Medium/long-range military and commercial freight transport.
Power Plant: Four 4,508 ehp Allison T56-A-15 turboprops.
Performance: Max. cruising speed, 386 mph (620 km/h) at 20,000 ft (6 095 m); long-range cruise, 345 mph (556 km/h); range (max. payload), 2,300 mls (3 700 km); ferry range (with 2,265 Imp gal/10 296 l of external fuel), 5,354 mls (8 617 km).
Weights: Operational empty, 79,516 lb (36 068 kg); max. take-off, 155,000 lb (70 310 kg).
Accommodation: Normal flight crew of four and provision for 97 casualty litters plus medical attendants, 128 combat troops or 92 paratroops. For pure freight role up to seven cargo pallets may be loaded.
Status: The L-100-30 and its military equivalent, the C-130H-30, are stretched versions of the basic Hercules, the C-130H. The original civil model, the L-100-20 featured a 100-in (2,54-m) fuselage stretch over the basic military model, and the L-100-30, intended for both military and civil application, embodies a further 80-in (2,03-m) stretch. Military operators of the C-130H-30 version are Algeria, Indonesia and Cameroun, and 30 of the RAF's Hercules C Mk 1s (equivalent of the C-130H) are being modified to C-130H-30 standards as Hercules C Mk 3s, 20 having been returned to service by the beginning of 1983 with the programme continuing into 1985. More than 1,670 Hercules (all versions) had been delivered by the beginning of 1983 when production was continuing at a rate of three monthly.
Notes: There are currently some 90 Hercules in commercial service, and numerous variants (eg, tanker, search and rescue, command communications) are serving with the US forces.

LOCKHEED L-100-30 HERCULES

Dimensions: Span, 132 ft 7 in (40,41 m); length, 112 ft 9 in (34,37 m); height, 38 ft 3 in (11,66 m); wing area, 1,745 sq ft (162,12 m²).

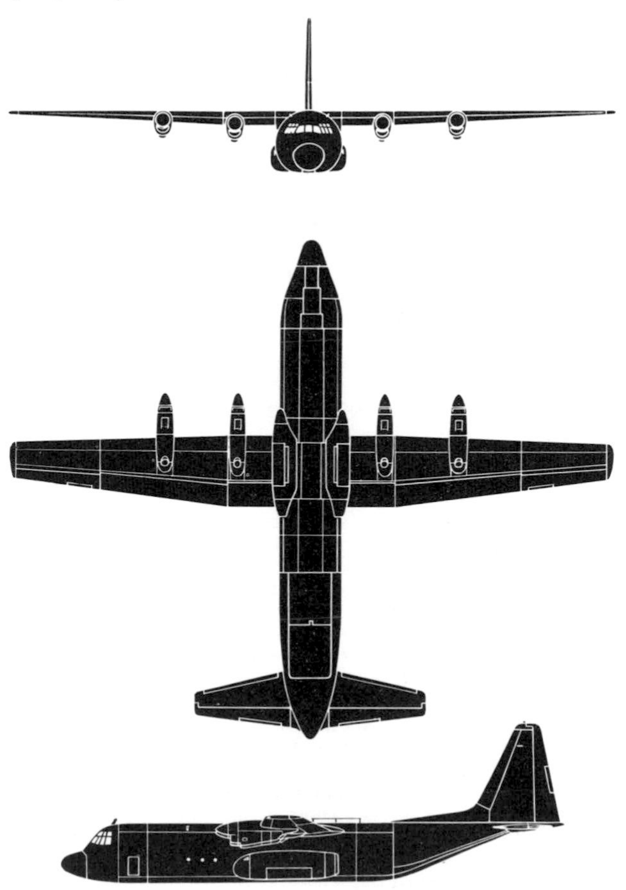

LOCKHEED TR-1

Country of Origin: USA.
Type: (TR-1A) Single-seat high-altitude tactical reconnaissance aircraft and (TR-1B) two-seat conversion trainer.
Power Plant: One 17,000 lb st (7 711 kgp) Pratt & Whitney J75-P-13B turbojet.
Performance: Max. cruising speed, 435 mph (700 km/h) at 70,000 ft (21 355 m); max. operational ceiling, 90,000 ft (27 430 m); max. range, 3,000 plus mls (4 830 plus km); max. endurance, 12 hrs.
Weights: Approx. max. take-off, 30,000 lb (13 608 kg).
Status: First TR-1A delivered in September 1981 against USAF procurement of 35 aircraft, including two two-seat TR-1Bs. First TR-1s to be assigned to USAF unit February 1983 with full 18-aircraft squadron by 1986. Current TR-1 production rate at beginning of 1983 four-five aircraft annually.
Notes: The TR-1 is a derivative of the U-2R and approximately 40 per cent larger than the original U-2, production of which terminated in 1968. Featuring updated sensors and introducing synthetic aperture radar, the TR-1A is intended to orbit 150 miles (240 km) or so behind the battlefield at an altitude of 60,000-70,000 ft (18 290-21 335 m) to provide reconnaissance information for tactical commanders, its sensors having a range in excess of 300 miles (480 km) and thus providing information relating to movements deep within enemy-held territory without crossing the forward edge of the battle area. The TR-1A has interchangeable noses, mission bay hatches and instrument wing pods, carrying nearly 4,000 lb (1 814 kg) of sensors, and may be fitted with an inflatable radome beneath the fuselage for the AWACS role.

LOCKHEED TR-1

Dimensions: Span, 103 ft 0 in (31,39 m); length, 63 ft 0 in (19,20 m); height, 16 ft 0 in (4,88 m).

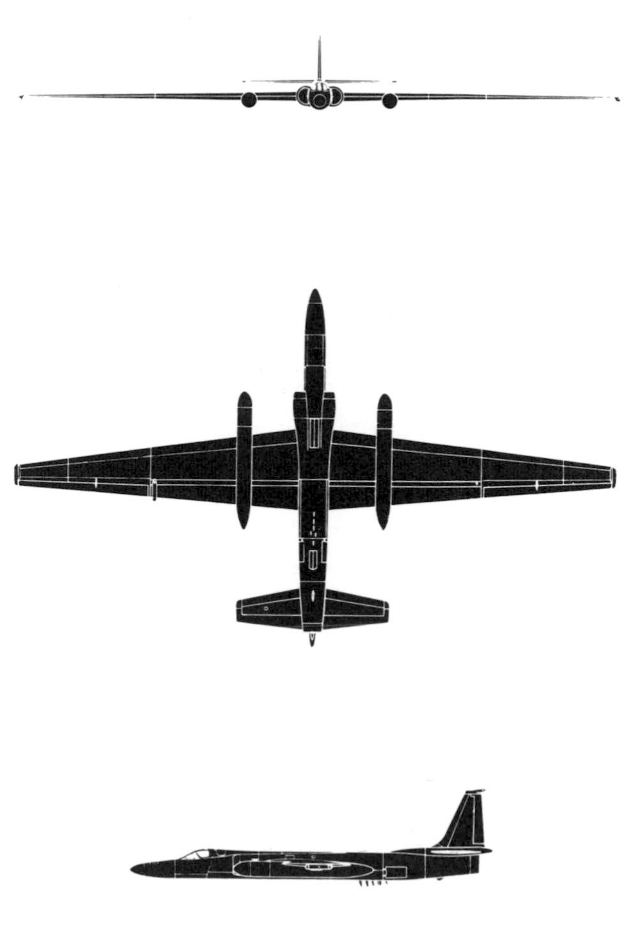

McDONNELL DOUGLAS (CAMMACORP) DC-8 SUPER 70 SERIES

Country of Origin: USA.
Type: Long-haul commercial airliner.
Power Plant: Four 24,000 lb st (10 900 kgp) General Electric/ SNECMA CFM56-1B turbofans.
Performance: Max. cruising speed, 583 mph (938 km/h) at 30,000 ft (9 145 m); econ cruise, 530 mph (854 km/h) at 35,000 ft (10 670 m); range cruise, 512 mph (825 km/h) at 35,000 ft (10 670 m); max. payload range (-71), 4,030 mls (6 486 km), (-72) 6,794 mls (10 933 km), (-73) 5,239 mls (8 432 km).
Weights: Max. take-off (-71), 325,000 lb (147 420 kg), (-72), 350,000 lb (158 760 kg), (-73), 355,000 lb (161 028 kg).
Accommodation: Flight crew of three and max. passenger capacity of (Srs 71 and 73) 269 or (Srs 72) 201.
Status: The Super 70 DC-8 is a modernised and re-engined version of the Super 60, production of which terminated in May 1972. The Srs 71, 72 and 73 are respectively conversions of the original Srs 61, 62 and 63, the programme being undertaken by McDonnell Douglas on behalf of Cammacorp (which company is responsible for marketing). Firm orders had been placed for almost 100 conversions by the beginning of 1983, with some 20 delivered and conversion rate proceeding at three monthly. The first Srs 71 flew on 15 August 1981, the first Srs 72 on 5 December 1981, and the first Srs 73 on 4 March 1982.
Notes: The DC-8 is the first large jet airliner to be subjected to a major re-engining programme, the Srs 71 version being illustrated both above and on the opposite page. The conversion offers a considerable reduction in fuel burn (15–17 per cent over the Srs 63 and 23 per cent over the Srs 61) and substantial improvements in take-off, climb and cruise altitude performance.

McDONNELL DOUGLAS (CAMMACORP) DC-8 SUPER 70 SERIES

Dimensions: Span (Srs 71), 142 ft 5 in (43,40 m), (Srs 72 and 73), 148 ft 5 in (45,23 m); length (Srs 71 and 73), 187 ft 5 in (57,12 m), (Srs 72), 157 ft 5 in (47,98 m); height, 43 ft 0 in (13,10 m); wing area (Srs 71), 2,884 sq ft (267,92 m²), (Srs 72 and 73), 2,927 sq ft (271,92 m²).

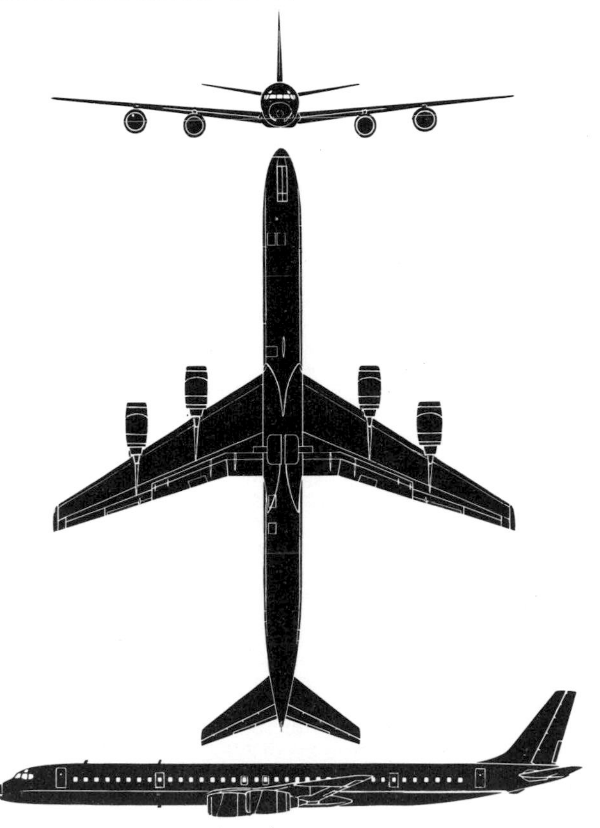

McDONNELL DOUGLAS DC-9 SUPER 80

Country of Origin: USA.
Type: Short/medium-haul commercial airliner.
Power Plant: Two 19,250 lb st (8 730 kgp) Pratt & Whitney JT8D-209 turbofans.
Performance: Max. cruising speed, 574 mph (924 km/h) at 27,000 ft (8 230 m); econ. cruise, 522 mph (840 km/h) at 33,000 ft (10 060 m); long-range cruise, 505 mph (813 km/h) at 35,000 ft (10 670 m); range (with max. payload), 1,594 mls (2 565 km) at econ cruise, (with max. fuel), 3,280 mls (5 280 km) at long-range cruise.
Weights: Operational empty, 77,797 lb (35 289 kg); max. take-off, 140,000 lb (63 503 kg).
Accommodation: Flight crew of two and typical mixed-class arrangement for 23 first- and 137 economy-class passengers, or 155 all-economy or 172 commuter-type arrangements with five-abreast seating.
Status: First Super 80 flown on 18 October 1979, with first customer delivery (to Swissair) on 12 September 1980, and 183 ordered by beginning of 1983 for 21 customers with 95 delivered. Total of 1,101 DC-9s of all versions ordered by beginning of 1983 (excluding aircraft being built for short-term lease) when production was 4·5 monthly.
Notes: The Super 80 is currently the largest of six members of the DC-9 family, the Super 82 sub-type having 20,850 lb st (9 458 kgp) JT8D-217 engines with which it was certificated at a max. take-off weight of 149,500 lb (67 813 kg) in September 1982, this giving a max. payload range of 2,300 mls (3 700 km). Current development studies include the Super 83 with a max. take-off weight of 160,000 lb (72 576 kg) and a max. payload range of the order of 2,880 mls (4 630 km). Overall size will remain unchanged, but wings and undercarriage will be strengthened to cater for the additional fuel and increased weights.

McDONNELL DOUGLAS DC-9 SUPER 80

Dimensions: Span, 107 ft 10 in (32,85 m); length, 147 ft 10 in (45,08 m); height, 29 ft 4 in (8,93 m); wing area, 1,279 sq ft (118,8 m²).

McDONNELL DOUGLAS DC-10 SERIES 30

Country of Origin: USA.
Type: Medium-haul commercial airliner.
Power Plant: Three 52,500 lb st (23 814 kgp) General Electric CF6-50C2 turbofans.
Performance: Max. cruising speed (at 396,830 lb/180 000 kg), 594 mph (956 km/h) at 31,000 ft (9 450 m); long-range cruise, 540 mph (870 km/h) at 31,000 ft (9 450 m); range (with max. payload), 4,856 mls (7 815 km), (with max. fuel), 6,300 mls (10 140 m).
Weights: Operational empty, 261,459 lb (118 597 kg); max. take-off, 572,000 lb (259 457 kg).
Accommodation: Flight crew of three and typical mix-class arrangements for 225-270 passengers. Maximum authorised single-class accommodation for 380 passengers.
Status: First DC-10 (Series 10) flown 29 August 1970, with first Series 30 (46th DC-10 built) following on 21 June 1972, this being preceded on 28 February 1972 by first Series 40. Orders (including KC-10A—see pages 134-5) totalled 382 at the beginning of 1983 with 376 delivered (including 12 KC-10As) and production (military and civil versions) continuing at one monthly.
Notes: The DC-10 Series 30 and 40 have identical fuselages to the Series 10 and 15, the former being intercontinental models differing in power plant, weight and wing details, and the latter being intended for domestic operation. The Series 40 differs from the Series 30 in having 53,000 lb st (24 040 kgp) Pratt & Whitney JT9D-59A turbofans. A lighter version of the DC-10 known as the MD-100 was under consideration at the beginning of 1983, possible engines being the Rolls-Royce RB. 211-535H4 or new-technology Pratt & Whitney 2000 series. Range of the MD-100 will be equivalent to that of the Series 30, with a 16% reduction in operating cost.

McDONNELL DOUGLAS DC-10 SERIES 30

Dimensions: Span, 165 ft 4 in (50,42 m): length, 181 ft 4¾ in (55,29 m); height, 58 ft 0 in (17,68 m); wing area, 3,958 sq ft (367,7 m²).

McDONNELL DOUGLAS AV-8B HARRIER II

Country of Origin: USA (and UK).
Type: Single-seat V/STOL ground attack aircraft.
Power Plant: One 21,180 lb st (9 607 kgp) Rolls-Royce F402-RR-406 Pegasus 11-21E (Mk 105) turbofan.
Performance: Max. speed (clean aircraft) 668 mph (1 075 km/h) or Mach 0·88 at sea level, 614 mph (988 km/h) or Mach 0·93 at 36,000 ft (10 970 m); tactical radius (HI-LO-HI interdiction with seven 1,000-lb/453,6-kg bombs and 25-mm cannon), 692 mls (1 114 km); ferry range (with four 250 Imp/gal/1 136 l drop tanks), 2,876 mls (4 630 km).
Weights: Operational empty, 12,750 lb (5 783 kg); max. take-off (for VTO), 19,185 lb (8 702 kg), (for STO), 29,750 lb (13 495 kg).
Armament: One 25-mm GAU-12/U five-barrel rotary cannon and up to 9,200 lb (4 173 kg) of ordnance on one fuselage centreline and six wing stations.
Status: First of four FSD (Full Scale Development) aircraft flown on 5 November 1981, with first of pilot batch of 12 aircraft to be delivered to US Marine Corps October 1983. The USMC has a requirement for 336 aircraft, and the RAF has an initial requirement for a further 60 (as Harrier GR Mk 5s), deliveries to the latter service being scheduled for 1986.
Notes: The AV-8B is a progressive development of the British Aerospace Harrier (see 1982 edition) currently serving with the RAF in GR Mk 3 and T Mk 4 forms, with USMC as the AV-8C and TAV-8A, and with the Spanish Navy as the Matador. The AV-8B Harrier II differs primarily in having a larger supercritical wing of composite construction, improved lift devices, extended rear fuselage and a raised cockpit. It has twice the payload or radius of action compared with the Harrier GR Mk 3.

McDONNELL DOUGLAS AV-8B HARRIER II

Dimensions: Span, 30 ft 4 in (9,24 m); length, 46 ft 4 in (14,12 m); height, 11 ft 8 in (3,55 m); wing area, 241 sq ft (22,40 m²).

McDONNELL DOUGLAS F-15C EAGLE

Country of Origin: USA.
Type: Single-seat air superiority fighter.
Power Plant: Two 14,780 lb st (6 705 kgp) dry and 23,904 lb st (10 855 kgp) reheat Pratt & Whitney F100-PW-100 turbofans.
Performance: Max. speed (short-endurance dash), 1,676 mph (2 698 km/h) or Mach 2·54, (sustained), 1,518 mph (2 443 km/h) or Mach 2·3 at 40,000 ft (12 190 m); max. endurance (internal fuel), 2·9 hrs, (with conformal pallets), 5·25 hrs; service ceiling, 63,000 ft (19 200 m).
Weights: Basic equipped, 28,700 lb (13 018 kg); loaded (full internal fuel and four AIM-7 AAMs), 44,500 lb (20 185 kg); max. take-off, 68,000 lb (30 845 kg).
Armament: One 20-mm M-61A1 rotary cannon plus four AIM-7F Sparrow and four AIM-9L Sidewinder AAMs.
Status: First flown 26 February 1979, the F-15C is the second major single-seat production version of the Eagle, having, together with its two-seat equivalent, the F-15D, supplanted the F-15A and F-15B from the 444th aircraft mid 1980. The F-15C and D were the current production models at the beginning of 1983, with 640 (all versions) delivered, total USAF Eagle requirement being 870 aircraft.
Notes: Featuring upgraded avionics and conformal fuel packs, the F-15C is being supplied to Saudi Arabia (47 plus 15 F-15Ds) and 84 are being licence manufactured as F-15Js by Japan (including eight from knocked-down assemblies) which country is also receiving 12 F-15DJ two-seaters. Israel received 40 F-15 Eagles (since modified for conformal tanks as seen on F-15D above) and is to receive a further 11 aircraft.

McDONNELL DOUGLAS F-15C EAGLE

Dimensions: Span, 42 ft 9¾ in (13,05 m); length, 63 ft 9 in (19,43 m); height, 18 ft 5½ in (5,63 m); wing area, 608 sq ft (56,50 m²).

McDONNELL DOUGLAS F-18A HORNET

Country of Origin: USA.
Type: Single-seat shipboard and shore-based multi-role fighter and attack aircraft.
Power Plant: Two 10,600 lb st (4 810 kgp) dry and 15,800 lb st (7 167 kgp) reheat General Electric F404-GE-400 turbofans.
Performance: Max. speed (AAMs on wingtip and fuselage stations), 1,190 mph (1 915 km/h) or Mach 1·8 at 40,000 ft (12 150 m); initial climb (half fuel and wingtip AAMs), 60,000 ft/min (304,6 m/sec); tactical radius (combat air patrol on internal fuel), 480 mls (770 km), (with three 262 Imp gal/ 1 192 l external tanks), 735 mls (1 180 km).
Weights: Empty equipped, 28,000 lb (12 700 kg); loaded (air superiority mission with half fuel and four AAMs), 35,800 lb (16 240 kg); max. take-off, 56,000 lb (25 400 kg).
Armament: One 20-mm M-61A-1 rotary cannon and (air-air) two AIM-7E/F Sparrow and two AIM-9G/H Sidewinder AAMs, or (attack) up to 17,000 lb (7 711 kg) of ordnance.
Status: First of 11 FSD (full-scale development) Hornets (including two TF-18A two-seaters) flown 18 November 1978. Planning at beginning of 1983 called for 1,366 Hornets for US Navy and US Marine Corps (including 153 TF-18As). First production F-18A flown April 1980.
Notes: Land-based versions of the Hornet have been ordered by Australia (57 F-18As and 18 TF-18As), Canada (113 CF-18As and 24 CF-18Bs) and Spain (84 F-18As and TF-18As). Separate F-18 fighter and A-18 attack versions of the Hornet were initially planned by the US Navy. Both roles were subsequently combined in a single basic version, and current planning calls for the inclusion of two Hornet squadrons in the complement of each of the large US Navy carriers.

McDONNELL DOUGLAS F-18A HORNET

Dimensions: Span, 37 ft 6 in (11,43 m); length, 56 ft 0 in (17,07 m); height, 15 ft 4 in (4,67 m); wing area, 396 sq ft (36,79 m²).

McDONNELL DOUGLAS KC-10A EXTENDER

Country of Origin: USA.
Type: Flight refuelling tanker and military freighter.
Power Plant: Three 52,500 lb st (23 814 kgp) General Electric CF6-50C2 turbofans.
Performance: Max. speed, 620 mph (988 km/h) at 33,000 ft (10 060 m); max. cruise, 595 mph (957 km/h) at 31,000 ft (9 450 m); long-range cruise, 540 mph (870 km/h); typical refuelling mission, 2,200 mls (3 540 km) from base with 200,000 lb (90 720 kg) of fuel and return; max. range (with 170,000 lb/77 112 kg freight), 4,370 mls (7 033 km).
Weights: Operational empty (tanker), 239,747 lb (108 749 kg), (cargo configuration), 243,973 lb (110 660 kg); max. take-off, 590,000 lb (267 624 kg).
Accommodation: Flight crew of five plus provision for six seats for additional crew and four bunks for crew rest. Fourteen further seats may be provided for support personnel in the forward cabin. Alternatively, a larger area can be provided for 55 more support personnel, with necessary facilities, to increase total accommodation (including flight crew) to 80.
Status: First KC-10A was flown on 12 July 1980, with 16 ordered by the USAF by the beginning of 1983. A further 44 are being ordered under five-year contracting process for delivery from 1983 through 1987. First operational KC-10A squadron was activated on 1 October 1981, and 12 had been delivered to the USAF by the beginning of 1983.
Notes: The KC-10A is a military tanker/freighter derivative of the commercial DC-10 Series 30 (see pages 126-7) with refuelling boom, boom operator's station, hose and drogue, military avionics and body fuel cells in the lower cargo compartments. The KC-10A is the first tanker developed from the outset to offer both flying-boom and probe-and-drogue refuelling.

McDONNELL DOUGLAS KC-10A EXTENDER

Dimensions: Span, 165 ft 4 in (50,42 m); length, 182 ft 0 in (55,47 m); height, 58 ft 1 in (17,70 m); wing area, 3,958 sq ft (367,7 m²).

MIKOYAN MIG-23 (FLOGGER)

Country of Origin: USSR.
Type: Single-seat (Flogger-B, E and G) air superiority and (Flogger-F and H) close air support fighter.
Power Plant: One 17,635 lb st (8 000 kgp) dry and 25,350 lb st (11 500 kgp) reheat Tumansky R-29B turbofan.
Performance: (Flogger-G) Max. speed (clean aircraft with half fuel), 1,520 mph (2 446 km/h) or Mach 2·3 above 36,090 ft (11 000 m); combat radius (high-altitude air-air mission with four AAMs), 530 mls (850 km), (with centreline combat tank), 700 mls.
Weights: Normal loaded (clean), 34,170 lb (15 500 kg); max. take-off, 44,312 lb (20 100 kg).
Armament: One 23-mm twin-barrel GSh-23L cannon and (B and G) two AA-7 Apex and two AA-8 Aphid, or (E) four AA-2-2 Advanced Atoll AAMs, or (F and H) up to 9,920 lb (4 500 kg) of bombs and missiles.
Status: Aerodynamic prototype of MiG-23 flown winter 1966-67, with service debut (Flogger-B) following 1971. Production rate of 50 (all versions) monthly continuing at beginning of 1983, when some 2,500 were estimated to be in Soviet service.
Notes: The Flogger-G (illustrated) is the latest air-air version of the MiG-23 in Soviet service, this being an improved variant of the Flogger-B (see 1982 edition) with revised vertical tail and other changes. Flogger-E is an export equivalent of Flogger-B, and Flogger-F and H are air-ground versions with redesigned forward fuselage essentially similar to that of the MiG-27 (see pages 140-1).

MIKOYAN MIG-23 (FLOGGER)

Dimensions: (Estimated) Span (17 deg sweep), 46 ft 9 in (14,25 m), (72 deg sweep), 27 ft 6 in (8,38 m); length (including probe), 55 ft 1½ in (16,80 m); wing area, 293·4 sq ft (27,26 m²).

MIKOYAN MIG-25 (FOXBAT)

Country of Origin: USSR.

Type: Single-seat (Foxbat-A) interceptor fighter and (Foxbat-B and -D) high-altitude reconnaissance aircraft.

Power Plant: Two 20,500 lb st (9 300 kgp) dry and 27,120 lb st (12 300 kgp) reheat Tumansky R-31 turbojets.

Performance: (Foxbat-A) Max. speed (short-period dash with four AAMs), 1,850 mph (2 980 km/h) or Mach 2·8 above 36,000 ft (10 970 m); max. speed at sea level, 650 mph (1 045 km/h) or Mach 0·85; initial climb, 40,950 ft/min (208 m/sec); service ceiling, 80,000 ft (24 385 m); combat radius (including allowance for Mach 2·5 intercept, 250 mls (400 km), (range optimised profile at econ power), 400 mls (645 km).

Weights: (Foxbat-A) Empty equipped, 44,100 lb (20 000 kg); max. take-off, 77,160 lb (35 000 kg).

Armament: Four AA-6 Acrid AAMs (two semi-active radar homing and two IR-homing).

Status: The MiG-25 entered service (in Foxbat-A form) in 1970, the photo/ELINT recce version (Foxbat-B) following in 1971 and the optimised ELINT version (Foxbat-D) in 1974.

Notes: The Foxbat-A (illustrated on opposite page) and Foxbat-C two-seat conversion trainer (illustrated above) have been exported to Algeria, Libya and Syria, and the latter, together with Foxbat-B, to India. An advanced two-seat interceptor derivative of the MiG-25, assigned the reporting name Foxhound, had been deployed by two Soviet regiments by the beginning of 1983, this being equipped with uprated engines, new lookdown-shootdown radar and AA-9 long-range radar-guided AAMs. The Foxbat-B and -D recce aircraft are reportedly capable of attaining Mach 3·2 in clean condition.

MIKOYAN MIG-25 (FOXBAT)

Dimensions: Span, 45 ft 9 in (13,94 m); length, 73 ft 2 in (22,30 m); height, 18 ft 4½ in (5,60 m); wing area, 602·8 sq ft (56,00 m²).

MIKOYAN MIG-27 (FLOGGER)

Country of Origin: USSR.

Type: Single-seat tactical strike and close air support fighter.

Power Plant: One 14,330 lb st (6 500 kgp) dry and 17,920 lb st (8 130 kgp) reheat Tumansky R-29-300 turbofan.

Performance: Max. speed (clean aircraft with half fuel), 685 mph (1 102 km/h) or Mach 0·95 at 1,000 ft (305 m), 1,056 mph (1 700 km/h) or Mach 1·6 at 36,090 ft (11 000 m); combat radius (HI-LO-HI mission profile on internal fuel with 4,410 lb/2 000 kg external ordnance), 310 mls (500 km).

Weights: (Estimated) Normal loaded (clean), 35,000 lb (15 875 kg); max. take-off, 45,000 lb (20 410 kg).

Armament: One 23-mm six-barrel rotary cannon and up to 7,716 lb (3 500 kg) of external ordnance on five stations.

Status: Evolved from the MiG-23 (see pages 136-7) as a dedicated air-ground aircraft, the MiG-27 is believed to have first entered service (in Flogger-D) form in 1975-76, with production continuing (in Flogger-J form) at the beginning of 1983.

Notes: Whereas the Flogger-F and H are minimum-change air-ground derivatives of the MiG-23, the MiG-27 has been tailored closely for ground attack. The forward fuselage is similar to that of the Flogger-F and H, apart from augmented side armour, but the rough-field undercarriage has necessitated bulging of the fuselage and a modified turbofan is installed with larger-area fixed intakes and shorter afterburner nozzle. The Flogger-J (illustrated) has a lenghtened nose and wing leading-edge extensions.

MIKOYAN MIG-27 (FLOGGER)

Dimensions: (Estimated) Span (17 deg sweep), 46 ft 9 in (14,25 m), (72 deg sweep), 27 ft 6 in (8,38 m); length, 54 ft 0 in (16,46 m); wing area, 293·4 sq ft (27,26 m²).

MITSUBISHI MU-300 DIAMOND I

Country of Origin: Japan.
Type: Light corporate transport.
Power Plant: Two 2,500 lb st (1 134 kgp) Pratt & Whitney (Canada) JT15D-4 turbofans.
Performance: Max. speed (at 11,000 lb/4 990 kg), 495 mph (796 km/h), (at 12,000 lb/5 443 kg), 501 mph (806 km/h) at 30,000 ft (9 150 m); typical cruise, 466 mph (750 km/h) at 39,000 ft (11 890 m); long-range cruise, 432 mph (695 km/h); initial climb, 3,100 ft/min (15,7 m/sec); max. operating altitude, 41,000 ft (12 495 m); max. range (with four passengers and VFR reserves), 1,738 mls (2 797 km), (IFR reserves), 1,450 mls (2 333 km).
Weights: Empty equipped, 9,100 lb (4 128 kg); max. take-off, 14,630 lb (6 636 kg).
Accommodation: Pilot and co-pilot/passenger on flight deck and various arrangements for up to nine passengers in main cabin. The standard eight-seat interior consists of a three-seat couch, four single seats in fore-and-aft pairs and a single seat behind the aft cabin divider.
Status: First of two prototypes flown on 29 August 1978, and first production aircraft flown 21 May 1981. First customer delivery on 27 September 1982. Production rate of eight monthly to be attained during 1983.
Notes: Wholly conceived by Mitsubishi, with basic manufacture taking place in Japan at Nagoya, final assembly of the Diamond taking place at the Mitsubishi facility at San Angelo, Texas, where systems and avionics are installed, and functional and flight testing performed.

MITSUBISHI MU-300 DIAMOND I

Dimensions: Span, 43 ft 5 in (13,23 m); length, 48 ft 4 in (14,73 m); height, 13 ft 9 in (4,19 m); wing area, 241·4 sq ft (22,43 m²).

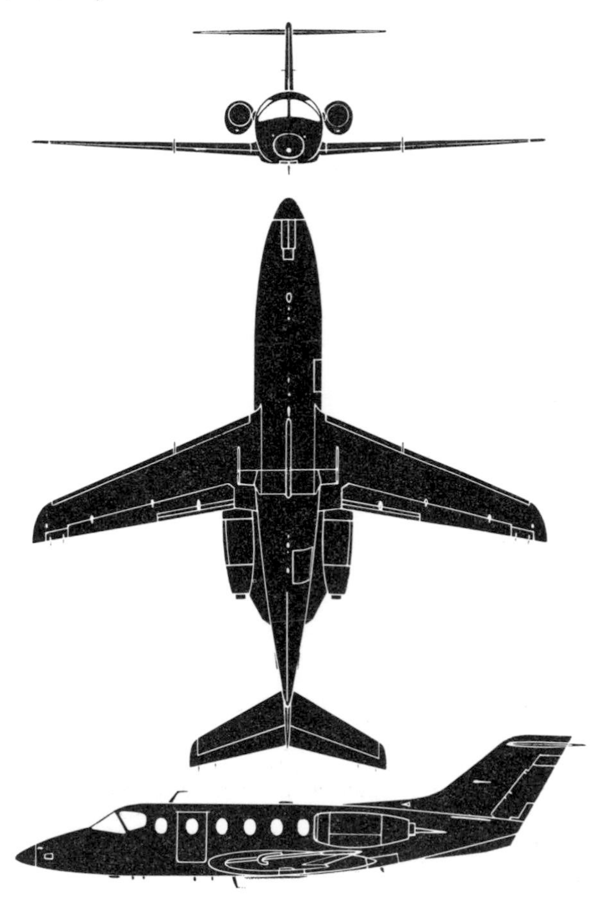

MUDRY CAP X

Country of Origin: France.
Type: Side-by-side two-seat primary trainer.
Power Plant: One 80 hp Mudry-Buchoux MB 4-80 four-cylinder horizontally-opposed engine.
Performance: (Estimated) Cruising speed (at 75% power), 112 mph (180 km/h) at 4,920 ft (1 500 m); range (at econ cruise), 497 mls (800 km).
Weights: Empty, 661 lb (300 kg); max. take-off, 1,168 lb (530 kg).
Status: The first prototype CAP X entered flight test on 10 September 1982, and current planning calls for initiation of series production during the course of 1983.
Notes: Designed by Auguste Mudry, the CAP X has been evolved to achieve the lowest possible initial cost combined with very low fuel consumption and minimal maintenance costs. The CAP X depends upon simplified production techniques utilising composite materials, large-scale production and an entirely new engine (the MB 4-80). Provision has been made in the design of the CAP X for a tailwheel undercarriage if preferred to the standard nosewheel undercarriage, and the first prototype is aerodynamically representative of the series model, but is of wooden construction throughout to facilitate the introduction of modifications should such be found necessary as a result of the flight test programme. The series model will incorporate composite materials in several components, including the wing spars, major fuselage longerons, some secondary structure, fairings and the moving control surfaces. These composites will include carbon fibre for the load-bearing members and honeycomb sandwich for the control surfaces. The design limitations are $+4{\cdot}4g$ to $-1{\cdot}8g$.

MUDRY CAP X

Dimensions: Span, 26 ft 3 in (8,00 m); length, 19 ft 4 in (5,90 m); height, 6 ft 9 in (2,05 m); wing area, 96·9 sq ft (9,00 m²).

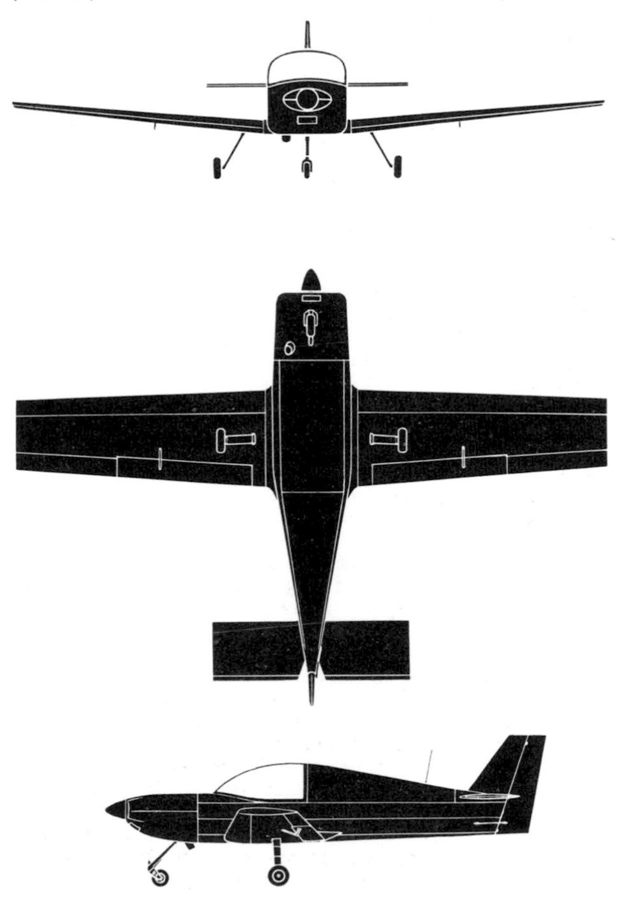

NDN-1T TURBO FIRECRACKER

Country of Origin: United Kingdom.
Type: Tandem two-seat basic trainer.
Power Plant: One 550 shp Pratt & Whitney (Canada) PT6A-25A turboprop.
Performance: (Provisional) Max. speed, 274 mph (441 km/h) at 17,000 ft (5 180 m); initial climb, 2,680 ft/min (13,6 m/sec); range (internal fuel and no reserves), 472 mls (760 km).
Weights: Empty equipped, 2,300 lb (1 043 kg); normal max. take-off, 3,250 lb (1 474 kg).
Armament: (Training or light strike) Four wing hardpoints permit various external ordnance loads (eg, four vertical tiers of four 81-mm Oerlikon SURA-D rockets).
Status: Initial order for three aircraft and options on a further four aircraft placed 1982 (by Specialist Flying Training) with deliveries commencing 1983.
Notes: The NDN-1T is a turboprop-powered version of the NDN-1 (see 1978 edition), which, powered by a 260 hp Avco Lycoming AEIO-540-B4D5 six-cylinder horizontally-opposed engine, entered test on 26 May 1977 and is illustrated above. The launch customer, Specialist Flying Training, provides flying training services for overseas governments. The NDN-1T, which is being manufactured by Firecracker Aircraft UK Limited, has been designed to simulate the handling feel and roll yaw characteristics of pure-jet swept and delta wing combat aircraft. To facilitate licence manufacture without sophisticated plant and equipment, the structural design of the NDN-1T includes no forgings or extrusions, and all parts may be cold-formed by hand during construction. Both seats are semi-reclined to increase the *g* tolerance of their occupants.

NDN-1T TURBO FIRECRACKER

Dimensions: Span, 26 ft 0 in (7,92 m): length, 28 ft 10½ in (8,80 m); height, 10 ft 0 in (3,05 m); wing area, 126 sq ft (11,71 m²).

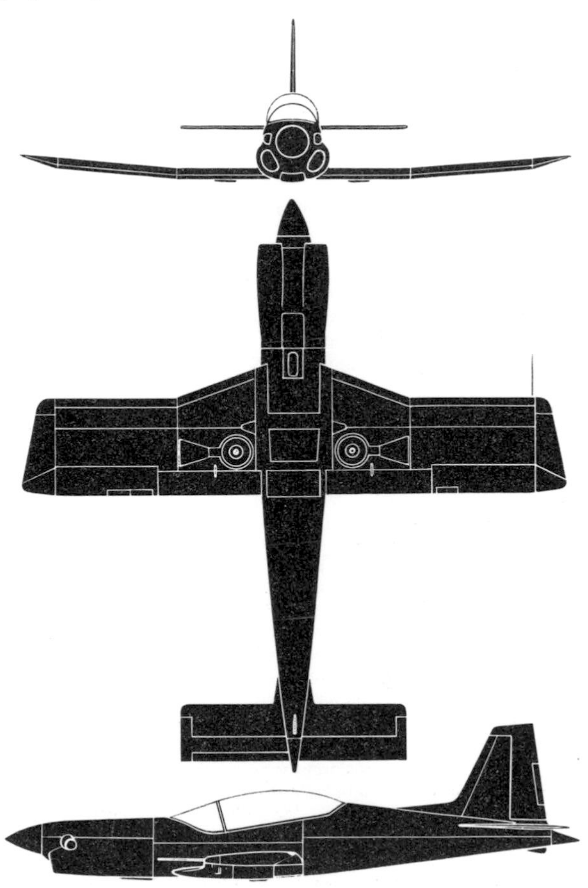

NDN-6 FIELDMASTER

Country of Origin: United Kingdom.
Type: Two-seat agricultural aircraft.
Power Plant: One 750 shp Pratt & Whitney (Canada) PT6A-34AG turboprop.
Performance: Max. speed (clean), 157 mph (252 km/h) at 5,500 ft (1 675 m); max. cruise, 155 mph (250 km/h) at sea level; initial climb (at 10,000 lb/4 536 kg), 600 ft/min (3,05 m/sec); range (max. fuel and two crew), 806 mls (1 297 km).
Weights: Empty equipped (typical), 4,500 lb (2 041 kg); max. take-off, 10,000 lb (4 536 kg).
Status: Prototype NDN-6 flown on 17 December 1981. Series production is scheduled to commence in 1984, high technology components (eg, the titanium hopper, heavy duty undercarriage, etc) being manufactured in the UK and then shipped with kits of materials for assembly overseas under an international co-production agreement.
Notes: The NDN-6 is claimed to be the first agricultural aircraft designed from the outset for turboprop power (ie, without a preceding piston-engined model) and will normally be flown as a single-seater, but the cockpit provides accommodation for a second person (eg, mechanic/loader, pupil pilot or fire spotter). The integral titanium hopper/tank (which has a capacity of 581 Imp gal/2 642 l) is part of the NDN-6's primary structure and carries the engine bearers at its forward end and the rear fuselage with cockpit aft. The full-span aerofoil flap incorporates a liquid spray dispersal system.

NDN-6 FIELDMASTER

Dimensions: Span, 50 ft 3 in (15,32 m); length, 36 ft 2 in (10,97 m); height, 12 ft 3 in (3,73 m); wing area, 338 sq ft (31,42 m²).

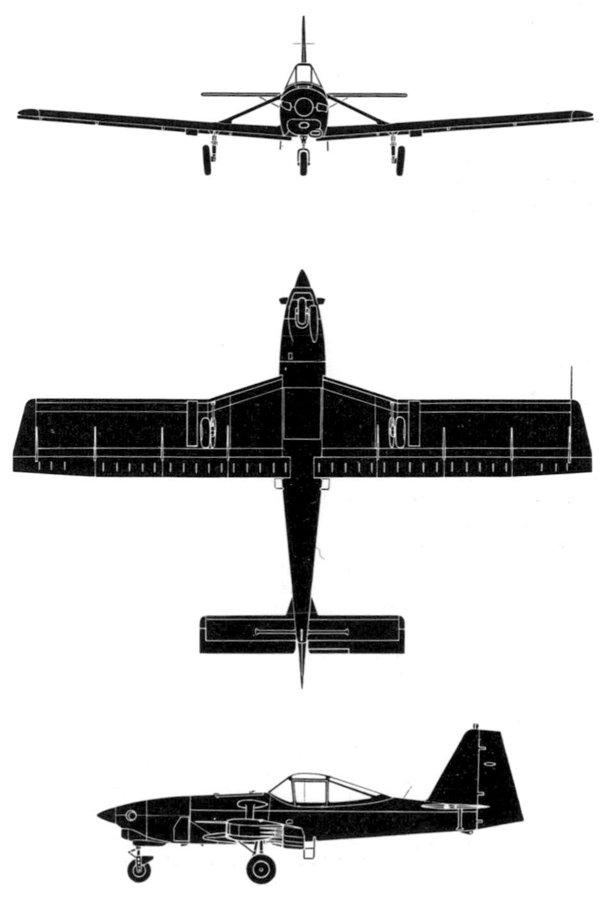

NORTHROP RF-5E TIGEREYE

Country of Origin: USA.
Type: Single-seat tactical reconnaissance aircraft.
Power Plant: Two 3,500 lb st (1 588 kgp) dry and 5,000 lb st (2 268 kgp) reheat General Electric J85-GE-21 turbojets.
Performance: Max. speed (clean configuration and 50% internal fuel), 843 mph (1 357 km/h) or Mach 1·17 at sea level, 997 mph (1 605 km/h) or Mach 1·51 at 36,090 ft (11 000 m); initial climb, 28,536 ft/min (145 m/sec); tactical radius (with one 220 Imp gal/1 000 l centreline drop tank and two AIM-9 Sidewinder AAMs), 287 mls (463 km) LO-LO-LO, 471 mls (759 km) HI-LO-HI, 569 mls (916 km) HI-HI-HI.
Weights: Empty, 9,750 lb (4 423 kg); loaded (clean), 15,550 lb (7 053 kg); max. take-off, 24,675 lb (11 192 kg).
Armament: One 20-mm M-39 cannon and two wingtip-mounted AIM-9 Sidewinder AAMs.
Status: Prototype RF-5E (adapted F-5E) flown 29 January 1979, and first production Tigereye flown on 15 December 1982 (first of two ordered for Royal Malaysian Air Force).
Notes: The RF-5E Tigereye is a dedicated tactical reconnaissance derivative of the F-5E Tiger II air superiority fighter (see 1981 edition) with a completely redesigned forward fuselage section to accommodate reconnaissance equipment. Provision is made for interchangeable sensor and camera pallets for different types of reconnaissance missions. One pallet has panoramic cameras and infrared line scanner for day and night missions up to 25,000 ft (7 620 m), another for day missions up to 50,000 ft (15 240 m), and a third for long-range oblique photography.

NORTHROP RF-5E TIGEREYE

Dimensions: Span (without missiles), 26 ft 8 in (8,13 m); length, 48 ft 10 in (14,88 m); height, 13 ft 4 in (4,06 m); wing area, 186 sq ft (17,30 m²).

NORTHROP F-20A TIGERSHARK

Country of Origin: USA.
Type: Single-seat multi-role fighter.
Power Plant: One 16,390 lb st (7 433 kgp) reheat General Electric F404-GE-F1G1 turbofan.
Performance: Max. speed, 1,320 mph (2 124 km/h) or Mach 2·0 above 36,000 ft (10 975 m), 800 mph (1 288 km/h) or Mach 1·05 at sea level; initial climb at combat weight (50% internal fuel and wingtip missiles), 54,100 ft/min (274,8 m/sec); combat ceiling, 53,050 ft (16 170 m); time to 40,000 ft (12 190 m) from brakes release, 2·2 min; tactical radius with two 229 Imp gal/1 040 l drop tanks and 20 min reserve at sea level (HI-LO-HI interdiction with seven Mk 82 bombs), 437 mls (704 km), (combat air patrol with 96 min on station), 345 mls (555 km); ferry range (max. external fuel), 1,842 mls (2 965 km).
Weights: Take-off (clean), 17,260 lb (7 829 kg); max. take-off, 26,290 lb (11 925 kg).
Armament: Two 20-mm M-39 cannon and up to 7,000 lb (3 175 kg) of external ordnance on five stations.
Status: First of four flight test and demonstration Tigersharks flown on 30 August 1982. Preparations for series production in hand at beginning of 1983, with deliveries offered 18-24 months of receipt of order.
Notes: The Tigershark is an advanced derivative of the F-5E Tiger II (see 1981 edition) with a low-bypass turbofan affording 70 per cent more thrust than the twin engines of the earlier fighter, integrated digital avionics, including a digital flight control system, and a multi-mode coherent pulse-Doppler radar. By comparison with the Tiger II, acceleration is enhanced 48 per cent, climb is increased by 32 per cent and maximum speed is raised by 38 per cent.

NORTHROP F-20A TIGERSHARK

Dimensions: Span, 26 ft 8 in (8,13 m); length, 46 ft 6 in (14,17 m); height, 13 ft 10 in (4,22 m); wing area, 186 sq ft (17,28 m²).

PANAVIA TORNADO F MK 2

Country of Origin: United Kingdom.
Type: Tandem two-seat interceptor fighter.
Power Plant: Two (approx) 9,000 lb st (4 082 kgp) dry and 16,000 lb st (7 258 kgp) reheat Turbo-Union RB.199-34R-04 Mk 101 (Improved) turbofans.
Performance: (Estimated) Max. speed, 920 mph (1 480 km/h) or Mach 1·2 at sea level, 1,450 mph (2 333 km/h) or Mach 2·2 at 40,000 ft (12 190 m); radius of action (combat air patrol with two 330 Imp gal/1 500 l drop tanks and allowance for two hours loiter), 350–450 mls (560–725 km); time to 30,000 ft (9 145 m) from brakes release, 1·7 min; ferry range, 2,650 mls (4 265 km).
Weights: (Estimated) Empty equipped, 25,000 lb (11 340 kg); max. take-off, 52,000 lb (23 587 kg).
Armament: One 27-mm IWKA-Mauser cannon, plus two AIM-9L Sidewinder and four BAe Sky Flash AAMs.
Status: First of three Tornado F Mk 2 prototypes flown on 27 October 1979, with production deliveries of 165–185 for the RAF commencing late 1983 and initial operational capability late 1984.
Notes: The F Mk 2 is the UK-only derivatives of the multinational (UK, Federal Germany and Italy) multi-role fighter (see 1978 edition) which entered operational service with the RAF and the Federal German *Luftwaffe* and *Marineflieger* during 1982. Retaining 80 per cent commonality with the multi-role version, the F Mk 2 features a redesigned nose for the intercept radar and a lengthened fuselage which increases internal fuel capacity and permits the mounting of four Sky Flash missiles on fuselage stations. Emphasis is placed on range and endurance in order to mount combat air patrols at considerable distances from the British coastline, and the aircraft is fitted with a permanently-installed retractable air-refuelling probe.

PANAVIA TORNADO F MK 2

Dimensions: Span (25 deg sweep), 45 ft $7\frac{1}{4}$ in (13,90 m), (68 deg sweep), 28 ft $2\frac{1}{2}$ in (8,59 m); length, 59 ft 3 in (18,06 m); height, 18 ft $8\frac{1}{2}$ in (5,70 m); wing area, 322·9 sq ft (30,00 m²).

PILATUS PC-7 TURBO TRAINER

Country of Origin: Switzerland.
Type: Tandem two-seat basic trainer.
Power Plant: One 550 shp Pratt & Whitney (Canada) PT6A-25A turboprop.
Performance: Max. continuous speed, 255 mph (411 km/h) at 10,000 ft (3 050 m); econ cruise, 230 mph (370 km/h) at 20,000 ft (6 100 m); initial climb, 2,065 ft/min (10,4 m/sec); service ceiling, 31,000 ft (9 450 m); max. range (internal fuel), 650 mls (1 047 km).
Weights: Empty equipped, 2,932 lb (1 330 kg); max. aerobatic, 4,188 lb (1 900 kg); max. take-off, 5,952 lb (2 700 kg).
Armament: (Training or light strike) Six wing hardpoints permit external loads up to max. of 2,292 lb (1 040 kg).
Status: First of two PC-7 prototypes flown 12 April 1966, and first production example flown 18 August 1978, with first customer deliveries (to Burma) following early 1979. The 200th PC-7 was rolled out on 30 July 1982, and total sales exceeded 300 aircraft by the beginning of 1983, with production continuing at a rate of six-seven monthly.
Notes: The PC-7 is a derivative of the piston-engined P-3, and has been selected by 10 air forces and naval air arms as follows: Angola (12), Abu Dhabi (14), Bolivia (36), Burma (17), Chile (10), Guatemala (12), Iraq (52), Malaysia (44), Mexico (55) and Switzerland (40). The PC-7 has also been adopted by the Swissair airline pilot training school (as illustrated above) and the Centre de Formation Aéronautique in France.

PILATUS PC-7 TURBO TRAINER

Dimensions: Span, 34 ft 1½ in (10,40 m); length, 31 ft 11⅞ in (9,75 m); height, 10 ft 6⅓ in (3,21 m); wing area, 178·68 sq ft (16,60 m²).

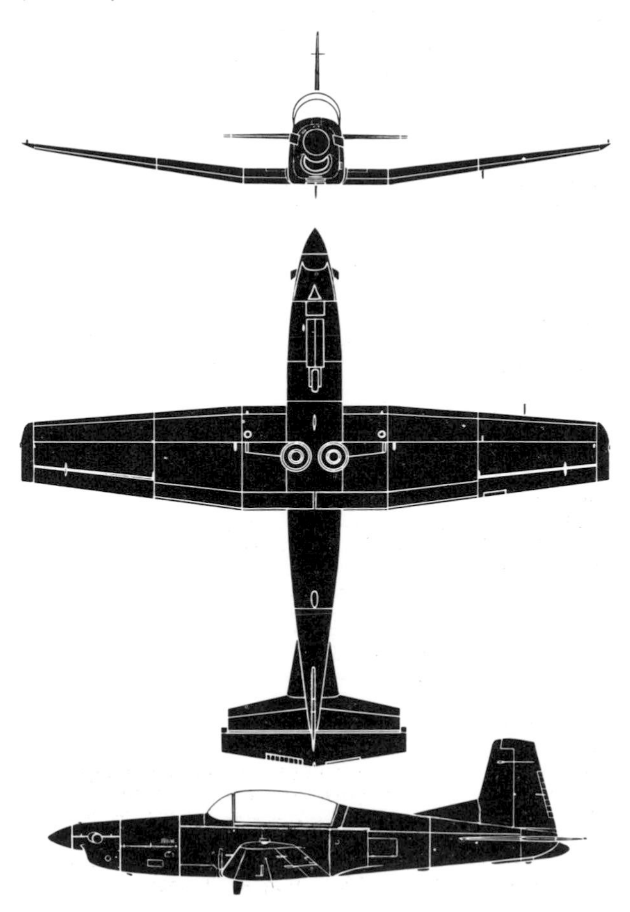

PIPER PA-31P-350 MOJAVE

Country of Origin: USA.
Type: Light cabin monoplane.
Power Plant: Two 350 hp Avco-Lycoming LTIO-540-V2AD six-cylinder horizontally-opposed engines.
Performance: Max. speed, 278 mph (448 km/h); cruise at optimum altitude (75% power), 269 mph (433 km/h), (at 65% power), 253 mph (407 km/h), (at 55% power), 231 mph (372 km/h); initial climb, 1,490 ft/min (7,57 m/sec); service ceiling, 30,400 ft (9 265 m); range, 1,370 mls (2 204 km) at 75% power, 1,444 mls (2 324 km) at 65% power, 1,496 mls (2 408 km).
Weights: Standard empty, 4,729 lb (2 155 kg); max. take-off, 7,200 lb (3 266 kg).
Accommodation: Side-by-side seats on flight deck with dual controls for pilot and co-pilot/passenger, and five individual seats in main cabin arranged as two facing pairs with central aisle and one inward facing seat at rear.
Status: Introduced late 1982, the Mojave is scheduled to be certificated in June 1983, with initial customer deliveries expected to follow within a few months.
Notes: The Mojave is a derivative design in that the fuselage is essentially similar to that of PA-31T-1 Cheyenne I, but with a lightened pressure vessel and an elongated wing generally similar to that of the PA-31-325 Navajo. The Mojave offers 860 lb (390 kg) of baggage capacity distributed between nose, aft cabin and engine nacelle compartments, and a 9,100 ft (2 775 m) cabin altitude can be maintained up to 25,000 ft (7 620 m), the planned maximum certified operating altitude.

PIPER PA-31P-350 MOJAVE

Dimensions: Span, 44 ft 6 in (13,56 m); length, 34 ft 6 in (10,51 m); jeight, 13 ft 0 in (3,96 m); wing area, 237 sq ft (22,02 m²).

PIPER PA-42 CHEYENNE IV

Country of Origin: USA.
Type: Light corporate executive transport.
Power Plant: Two 1,000 shp Garrett AiResearch TPE331-14 turboprops.
Performance: (Estimated) Max. cruising speed, 404 mph (650 km/h) at 22,500 ft (6 860 m); initial climb, 3,400 ft/min (17,27 m/sec); time to 35,000 ft (10 670 m), 14 min; service ceiling, 44,000 ft (13 410 m); range (eight passengers and IFR reserves), 1,360 mls (2 187 km); max. range (with 45 min reserves), 2,590 mls (4 170 km).
Weights: Empty, 7,050 lb (3 198 kg); max. take-off, 11,950 lb (5 420 kg).
Accommodation: Flight crew of one or two on separate flight deck and various optional arrangements for six to nine passengers in main cabin.
Status: First of three prototypes of the Cheyenne IV scheduled to commence flight test February 1983, with certification following early 1984 and initial customer deliveries in May of that year.
Notes: The Cheyenne IV is a re-engined derivative of the Cheyenne III (see 1982 edition) which is powered by 720 shp PT6A-41 turboprops. The airframe of the Cheyenne IV is basically similar to that of the III (illustrated above) and is externally little changed, but the fuselage structure has been strengthened to cater for a 10,000 ft (3 050 m) cabin pressure up to 41,000 ft (12 495 m) and the larger-diameter four-bladed Dowty Rotol composite propellers that replace the three-bladed Hartzell Q-tip propellers have necessitated adoption of a lengthened undercarriage. The engine nacelle extension baggage compartments of the Cheyenne III have been eliminated and the nose baggage compartment has been enlarged.

PIPER PA-42 CHEYENNE IV

Dimensions: Span, 47 ft $8\frac{1}{8}$ in (14,53 m); length, 43 ft $4\frac{3}{4}$ in (13,23 m); height, 15 ft 6 in (4,72 m); wing area, 293 sq ft (27,20 m²).

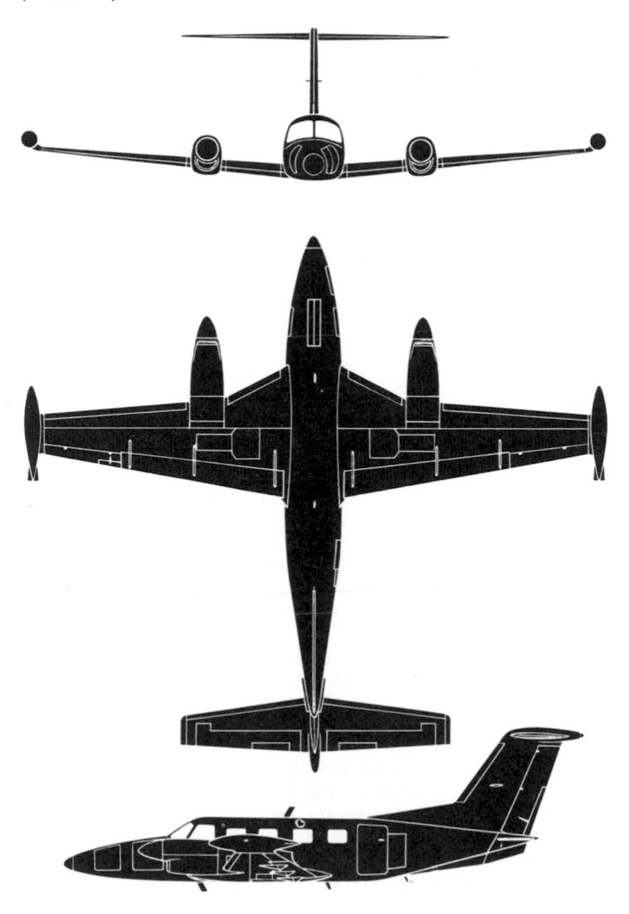

PIPER PA-46-310P MALIBU

Country of Origin: USA.
Type: Light cabin monoplane.
Power Plant: One 310 hp Continental TSIO-520-BE six-cylinder horizontally-opposed engine.
Performance: Max. speed, 254 mph (409 km/h) at optimum altitude; cruise (75% power), 239 mph (385 km/h), 65% power), 225 mph (363 km/h); initial climb, 1,143 ft/min (5,8 m/sec); range (with 45 min reserves), 1,542 mls (2 482 km) at 75% power, 1,657 mls (2 667 km) at 65% power, 1,830 mls (2 945 km) at 55% power.
Weights: Standard empty, 2,275 lb (1 032 kg); max. take-off, 3,850 lb (1 746 kg).
Accommodation: Pilot and five passengers in paired individual seats with rear airstair door.
Status: First prototype Malibu flown late 1980, with first production aircraft flying August 1982. Certification expected June 1983, with first delivery to a dealer following in August.
Notes: Intended to compete with the Cessna P210 Centurion, which, prior to the advent of the Malibu, was the sole pressurised single-engined cabin monoplane on the market, this new Piper aircraft is claimed to be the first production single-engined general aviation model to utilise computer-aided design and manufacturing (CAD/CAM) techniques. Possessing no relationship to previous Piper designs, the Malibu offers a cabin of "business twin" proportions with club seating and a rear airstair door, and is designed to give an 8,000 ft (2 440 m) cabin pressure up to 25,000 ft (7 620 m), at which the intercooling techniques of the TSIO-520 engine result in 240 hp still being available.

PIPER PA-46-310P MALIBU

Dimensions: Span, 43 ft 0 in (13,10 m); length, 28 ft $4\frac{3}{4}$ in (8,66 m); height, 11 ft $3\frac{1}{2}$ in (3,44 m); wing area, 175 sq ft (16,26 m²).

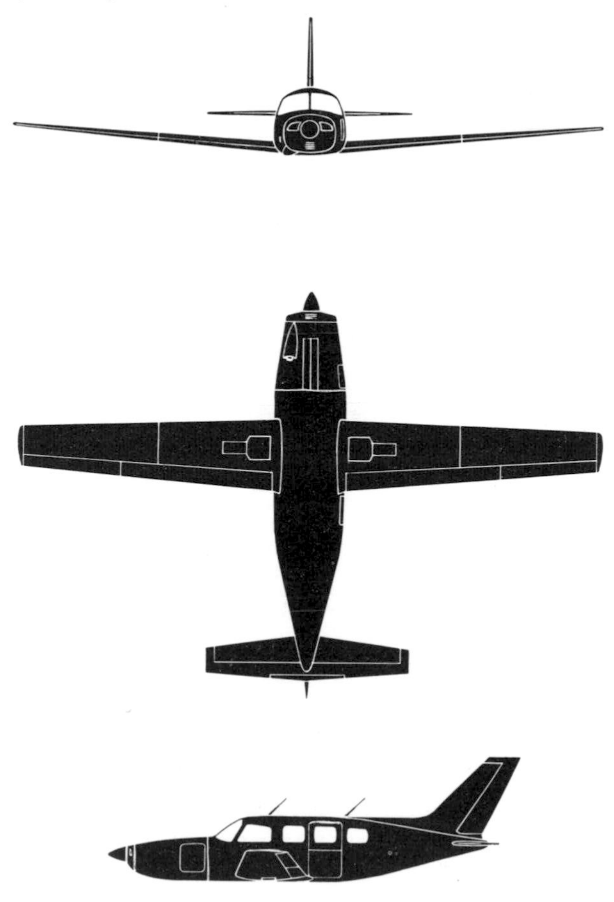

PIPER PA-48 ENFORCER

Country of Origin: USA.
Type: Single-seat close air support aircraft.
Power Plant: One 2,445 shp Avco Lycoming T55-L-9A turboprop.
Performance: (Estimated) Max. speed, 363 mph (584 km/h); tactical radius (with two 30-mm cannon pods), 460 mls (740 km).
Weights: Max. take-off, 14,000 lb (6 350 kg).
Armament: All ordnance is carried externally on six wing stations, weapons loads including two General Electric GPU-5 30-mm cannon pods, Canadian Bristol CRV-7 2·75-in (7-cm) rockets in nine- and 19-rocket pods, Mk 20 Rockeye and Mk 82 Snakeye rockets, and various selections of Minigun pods and cluster bombs.
Status: First of two PA-48 prototypes was scheduled to enter flight test late January 1983. Both prototypes are to participate in a flight test programme (to be conducted jointly by the manufacturer and the USAF) involving 293 sorties totalling some 390 flying hours and scheduled for completion in November 1983.
Notes: Built under a contract awarded by the USAF in September 1981, the PA-48 is based on the original Enforcer (illustrated above), the first of two prototypes of which was flown on 29 April 1971. The Enforcer was, in turn, based on the North American P-51 Mustang of WWII, but the new PA-48 retains less than 10 per cent component parts similar to the original P-51, is larger and heavier, makes extensive use of composite armour, has an ejection seat for the pilot, utilises a modified Gulfstream I main undercarriage, and has fixed 100 Imp gal (454 l) wingtip tanks.

PIPER PA-48 ENFORCER

Dimensions: Span, 41 ft 3½ in (12,59 m); length, 34 ft 2½ in (10,42 m); height, 13 ft 1¼ in (3,99 m); wing area, 257 sq ft (23,87 m²).

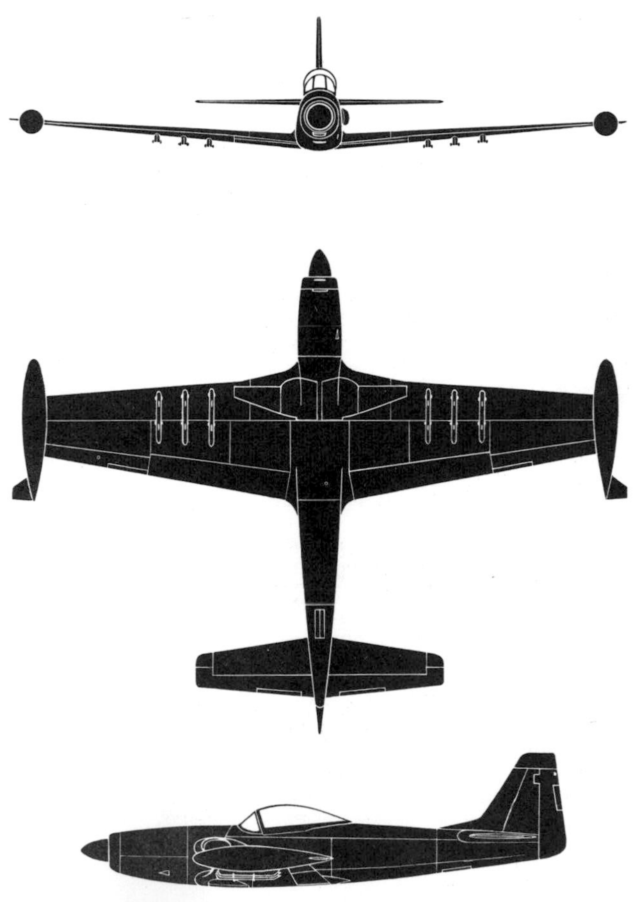

PIPER T-1040

Country of Origin: USA.
Type: Light regional airliner.
Power Plant: Two 500 shp Pratt & Whitney (Canada) PT6A-11 turboprops.
Performance: Max. speed, 280 mph (450 km/h); max. cruise, 234 mph (376 km/h) at sea level, 273 mph (439 km/h) at 11,000 ft (3 355 m); initial climb, 1,610 ft/min (8,18 m/sec); service ceiling, 27,900 ft (8 505 m); range (nine passengers and VFR reserves), 718 mls (1 156 km), (max. payload), 518 mls (834 km), (with 66 US gal/250 l wingtip tanks), 1,140 mls (1 835 km) plus 45 min reserves.
Weights: Basic, 5,114 lb (2 320 kg); max. take-off, 9,000 lb (4 082 kg).
Accommodation: Eleven individual seats in cabin, including one/two pilots' seats with full dual control.
Status: Prototype flown on 17 July 1981, with customer deliveries commencing July 1982.
Notes: The T-1040 is an optimised regional airliner utilising the basic wing, nose and tail of the PA-31T Cheyenne light corporate transport mated with the fuselage of the PA-31-350 Chieftain. The T-1040 is manufactured in parallel with the T-1020, these being the first two Piper aircraft evolved specifically for airline use and both affording similar capacity but the T-1020 being powered by 350 hp Avco Lycoming TIO/LTIO-540-J2BD "flat six" engines. Both types are aimed at the lower end of the regional airliner market and embody experience gained by small airlines with the PA-31-350 Chieftain. A 30 cu ft (0·85 m³) capacity ventral cargo pod was available from February 1983.

PIPER T-1040

Dimensions: Span, 41 ft 1 in (12,52 m); length, 36 ft 8 in (11,18 m); height, 12 ft 9 in (3,89 m); wing area, 229 sq ft (21,27 m²).

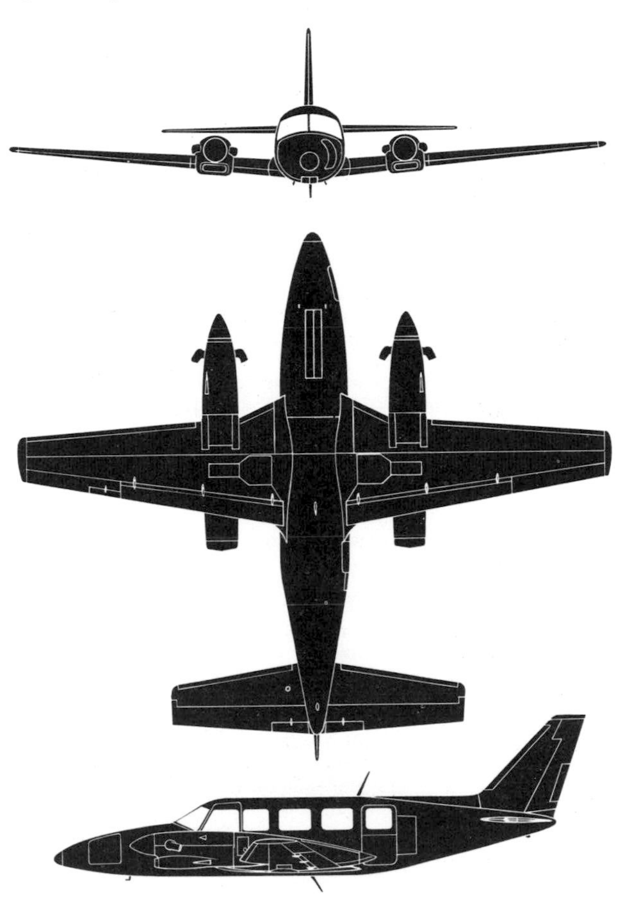

PZL M-21 MINI DROMADER

Country of Origin: Poland.
Type: Single-seat agricultural aircraft.
Power Plant: One 592 hp PZL-3SR seven-cylinder radial air-cooled engine.
Performance: Max. speed, 143 mph (230 km/h) at sea level; max. cruising speed, 141 mph (227 km/h); operating speed range, 96–112 mph (155–180 km/h); operating endurance (with 30-min allowance for transit), 3·6 hrs; max. range (no reserves), 435 mls (700 km).
Weights: Max. take-off, 7,275 lb (3 300 kg).
Status: The prototype M-21 entered flight test mid-1982, and is scheduled to enter production at the WSK-PZL Mielec factory in 1986, the anticipated production rate being 50 aircraft per year.
Notes: The latest in the PZL range of agricultural aircraft, the M-21 Mini Dromader (Dromedary) is essentially a scaled-down, lower-powered derivative of the M-18 Dromader (see 1981 edition) which it is intended to complement in service. The Mini Dromader possesses some 70 per cent commonality with the Dromader and has a payload of 1,984 lb (900 kg), the tank forward of the cockpit having a capacity of 374 Imp gal (1 700 l) of liquid chemical. The Mini Dromader is intended for the aerial application of water- and oil-based solutions or aerosols, seed sowing and crop dusting of areas not exceeding 50 hectares. Current planning involves manufacture of the Mini Dromader in five versions, including one powered by a turboprop.

PZL M-21 MINI DROMADER

Dimensions: Span, 47 ft 7¼ in (14,51 m); length, 31 ft 1¼ in (9,48 m); height, 12 ft 6⅓ in (3,82 m).

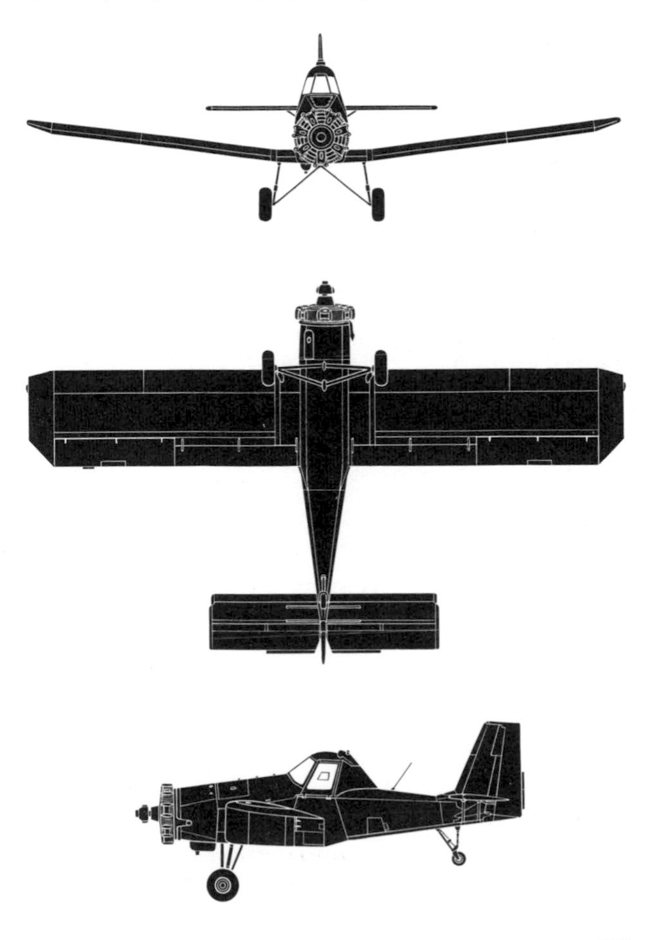

RHEIN-FLUGZEUGBAU FANTRAINER

Country of Origin: Federal Germany.
Type: Tandem two-seat primary/basic trainer.
Power Plant: Two (Fantrainer 400) 420 shp Allison 250-C20B or (Fantrainer 600) 600 shp Allison 250-C30 turboshafts driving five-bladed ducted fan.
Performance: (Fantrainer 400) Max. speed, 230 mph (370 km/h) at 10,000 ft (3 050 m); initial climb, 2,000 ft/min (10,2 m/sec); range (no reserves), 1,094 mls (1 760 km). (Fantrainer 600) Max. speed, 267 mph (430 km/h) at 18,000 ft (5 485 m); cruising speed, 230 mph (370 km/h) at 10,000 ft (3 050 m); initial climb, 3,150 ft/min (16 m/sec); range (no reserves), 864 mls (1 390 km).
Weights: (Fantrainer 400) Empty, 2,070 lb (939 kg); max. take-off, 3,484 lb (1 580 kg). (Fantrainer 600) Empty, 2,340 lb (1 060 kg); max. take-off, 5,070 lb (2 300 kg).
Status: First of two prototypes flown on 27 October 1977. Production deliveries to commence early 1984 against order for 47 (16 Fantrainer 400s and 31 600s) from Royal Thai Air Force of which first six to be completed by parent company and remainder to be assembled in Thailand from component sets supplied by parent company and Thai-manufactured wings.
Notes: Unconventional design intended to provide jet-type handling characteristics with low initial and operational costs. Whereas the wings of the prototypes are constructed mainly of glassfibre and plastics tube sandwich, the aircraft ordered by Thailand will have all-metal wings which will be offered as options to other potential customers.

RHEIN-FLUGZEUGBAU FANTRAINER

Dimensions: Span, 31 ft 10 in (9,70 m); length, 29 ft 6 in (9,00 m); height, 9 ft 6 in (2,90 m); wing area, 149·6 sq ft (13,90 m²).

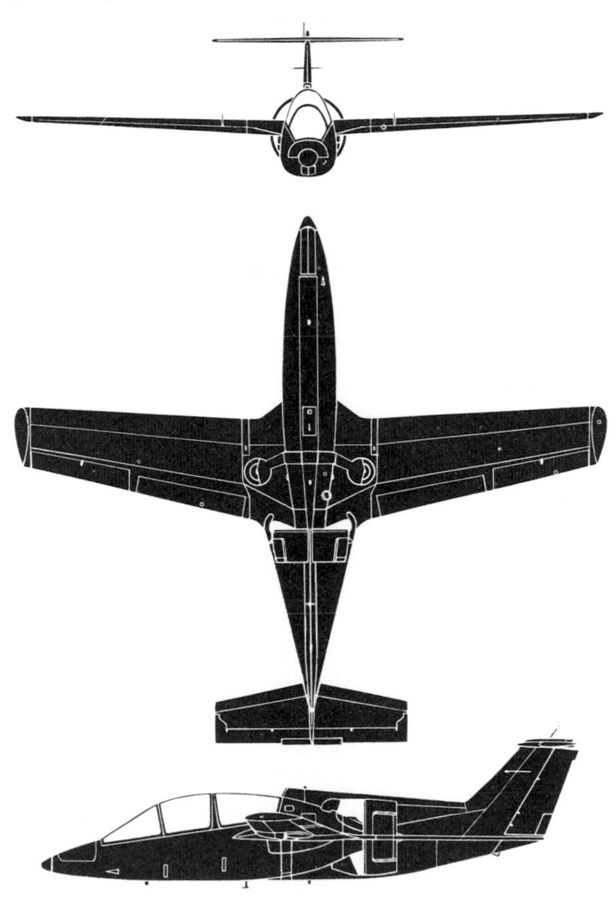

ROBIN R 3140

Country of Origin: France.
Type: Light cabin monoplane.
Power Plant: One 143 hp Avco Lycoming O-320-D2A four-cylinder horizontally-opposed engine.
Performance: Cruising speed (75% power), 153 mph (246 km/h) at 8,000 ft (2 400 m); econ cruise (65% power), 145 mph (234 km/h) at 11,000 ft (3 350 m); initial climb, 807 ft/min (4,1 m/sec); service ceiling, 14,500 ft (4 420 m); range (standard fuel), 516 mls (830 km), (max. fuel), 994 mls (1 600 km).
Weights: Empty, 1,268 lb (575 kg); max. take-off, 2,205 lb (1 000 kg).
Accommodation: Four seats in pairs.
Status: First prototype R 3140 flown on 8 December 1980, with second prototype with redesigned and definitive wing flying early 1982. Certification was expected by late 1982, with initial production deliveries during second quarter of 1983.
Notes: The R 3140 is the first in Avions Pierre Robin's R 3000 series of all-metal light aircraft which is to be marketed by SOCATA. Current planning calls for production of 50 R 3000 series aircraft during 1983, followed by 75 in 1984 and 100 in 1985. All the initial year's production is expected to be of the R 3140 version, but proposed variants utilising essentially the same structure are the R 3100 two-seater (100 hp O-235), the R 3120 two-plus-two seater (120 hp O-235), the R 3160R glider tug (160 hp O-360) and the R 3180S four-seater (180 hp TO-360) with turbo-supercharging and a retractable undercarriage.

ROBIN R 3140

Dimensions: Span, 29 ft 10¼ in (9,81 m); length, 24 ft 7½ in (7,51 m); height, 8 ft 8¾ in (2,66 m); wing area, 155·75 sq ft (14,47 m²).

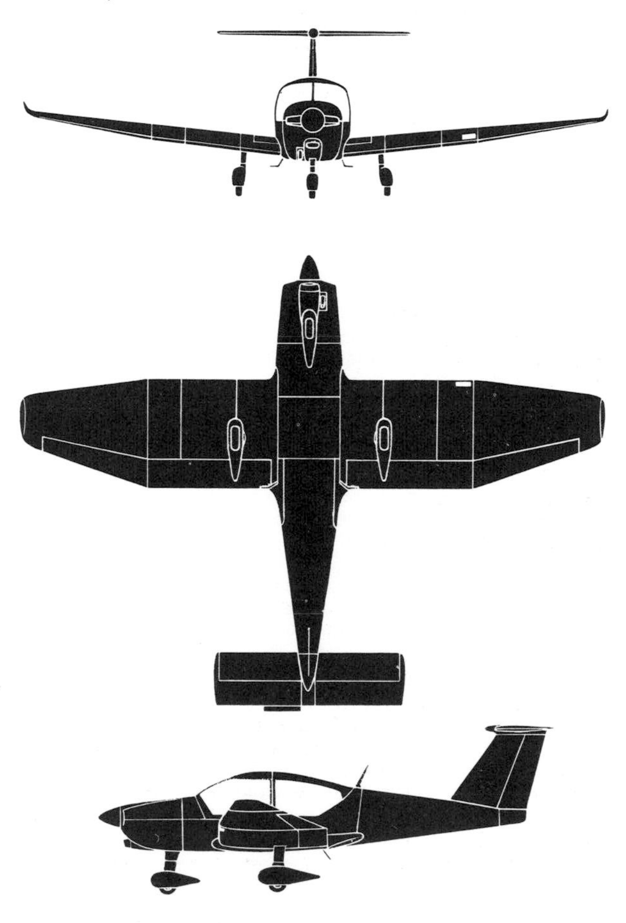

ROCKWELL B-1B

Country of Origin: USA.
Type: Strategic bomber and cruise missile carrier.
Power Plant: Four 30,750 lb st (13 948 kgp) General Electric F101-GE-102 turbofans.
Performance: Max. speed (clean condition), 792 mph (1 275 km/h) or Mach 1·2 at 40,000 ft (12 190 m); low-level penetration speed, 610 mph (980 km/h) or Mach 0·8.
Weights: Empty, 179,985 lb (81 641 kg); max. take-off, 477,000 lb (216 367 kg).
Accommodation: Flight crew of four comprising pilot, co-pilot and offensive and defensive systems operators.
Armament: Eight AGM-86B cruise missiles and 12 AGM-69 defence-suppression missiles internally, plus 12–14 AGM-86Bs externally, or 84 500-lb (227-kg) Mk 82 bombs internally, plus 44 externally, or 24 free-falling B-61 nuclear bombs, plus 14 externally.
Status: First contract placed 20 January 1982 in programme entailing manufacture of 100 B-1Bs, the first being scheduled for flight test late 1984/early 1985. Fifteenth B-1B to be delivered mid-1986, with production attaining four monthly by late 1986 and 100th delivered by April 1988.
Notes: The B-1B is a derivative of the Mach 2·2 B-1, first of four prototypes of which flew 23 December 1974. A 347-flight, 1,895-hour test programme was completed with these aircraft on 30 April 1981. The B-1B will have reduced speed capability by comparison with the B-1, being optimised for low-level penetration, the fourth B-1 (illustrated above) having been restored to flight test status in August 1982 to participate in an additional 1,100-hour test programme in which it will be joined mid-1983 by the second B-1. Together with the first production aircraft, these will complete the B-1B development.

ROCKWELL B-1B

Dimensions: Span (15 deg), 136 ft 8½ in (41,67 m), (67·5 deg), 78 ft 2½ in (23,84 m); length, 146 ft 8 in (44,70 m); height, 33 ft 7¼ in (10,24 m); wing area (approx), 1,950 sq ft (181,2 m²).

SAAB (JA) 37 VIGGEN

Country of Origin: Sweden.
Type: Single-seat all-weather interceptor fighter with secondary strike capability.
Power Plant: One 16,200 lb st (7 350 kgp) dry and 28,110 lb st (12 750 kgp) reheat Volvo Flygmotor RM 8B turbofan.
Performance: Max. speed (with four AAMs), 838 mph (1 350 km/h) or Mach 1·1 at sea level, 1,255–1,365 mph (2 020–2 195 km/h) or Mach 1·9–2·1 at 36,090 ft (11 000 m); time (from brakes off) to 32,810 ft (10 000 m), 1·4 min; tactical radius (Mach 2·0 interceptor mission), 250 mls (400 km), (counterair mission with centreline drop tank and 3,000 lb/ 1 360 kg of external ordnance), 650 mls (1 046 km) HI-LO-HI, 300 mls (480 km) LO-LO-LO.
Weights: Empty (approx), 26,895 lb (12 200 kg); combat (cannon armament and half fuel), 33,070 lb (15 000 kg), (with four AAMs), 37,040 lb (16 800 kg); max. take-off, 49,600 lb (22 500 kg).
Armament: One 30-mm Oerlikon KCA cannon and (intercept) two Rb 72 Sky Flash and two (or four) Rb 24 Sidewinder AAMs, or (interdiction) 13,227 lb (6 000 kg) of external ordnance.
Status: First of four JA 37 prototypes (modified from AJ 37 airframes) flown June 1974, with fifth and definitive prototype flown 15 December 1975. First production JA 37 flown on 4 November 1977, and total of 149 JA 37s (of 329 Viggens of all types) being produced for Swedish Air Force with some 60 delivered by beginning of 1983 and final deliveries scheduled for 1985.
Notes: The JA 37 is an optimised interceptor derivative of the AJ 37 attack aircraft (see 1973 edition).

SAAB (JA) 37 VIGGEN

Dimensions: Span, 34 ft 9¼ in (10,60 m); length (excluding probe) 50 ft 8¼ in (15,45 m); height, 19 ft 4¼ in (5,90 m); wing area (including foreplanes), 561·88 sq ft (52,20 m²).

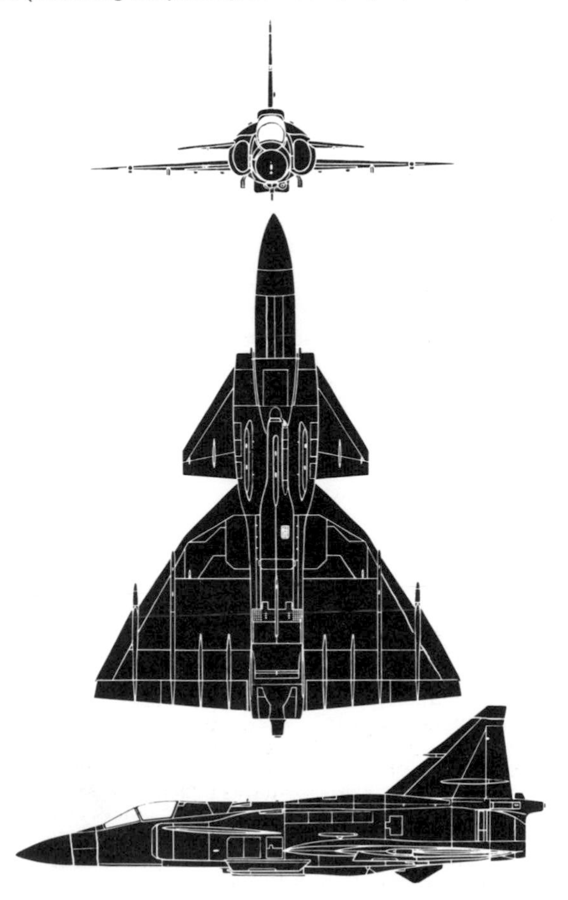

SAAB-FAIRCHILD 340

Countries of Origin: Sweden and USA.
Type: Regional airliner and corporate transport.
Power Plant: Two 1,675 shp General Electric CT7-5A or (corporate version) 1,600 shp CT7-7E turboprops.
Performance: Max. cruising speed, 315 mph (508 km/h) at 20,000 ft (6 095 m); long-range cruise, 243 mph (391 km/h) at 25,000 ft (7 620 m); max. cruise range (34 passengers), 1,036 mls (1 668 km), (22 passengers), 1,727 mls (2 780 km), (corporate version at long-range cruise with 2,400-lb/1 089-kg payload), 2,120 mls (3 410 km).
Weights: Typical operational empty, 15,860 lb (7 194 kg); max. take-off, 26,000 lb (11 794 kg).
Accommodation: Flight crew of two (with provision for third member) and standard regional airliner arrangement for 34 passengers three-abreast with offset aisle. Standard corporate arrangement providing 16 seats with various options.
Status: First prototype rolled out 27 October 1982 and scheduled to enter flight test January 1983, with second and third following in May and July 1983 respectively. First delivery (to Crossair) planned for April 1984, with total of 24 scheduled for delivery during course of that year. Production rate to peak at six monthly during 1987. Approximately 100 orders and options by beginning of 1983.
Notes: Subject of a joint programme between Saab-Scania (Sweden) and Fairchild Swearingen (USA) with development costs shared 65–35 between Swedish and US partners. Manufacture of the fuselage and final assembly is undertaken by Saab–Scania, the wing, tail surfaces and engine nacelles being produced by Fairchild Swearingen.

SAAB-FAIRCHILD 340

Dimensions: Span, 70 ft 4 in (21,44 m); length, 64 ft 8 in (19,72 m); height, 22 ft 6 in (6,87 m); wing area, 450 sq ft (41,80 m²).

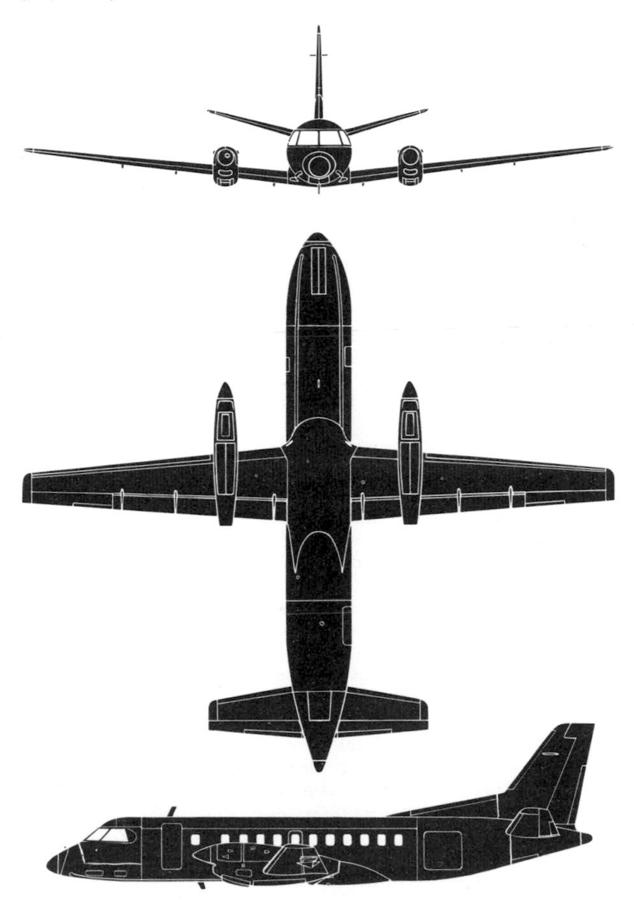

SEPECAT JAGUAR INTERNATIONAL

Countries of Origin: United Kingdom and France.
Type: Single-seat tactical strike fighter.
Power Plant: Two 5,520 lb st (2 504 kgp) dry and 8,400 lb st (3 811 kgp) reheat Rolls-Royce/Turboméca RT172-58 Adour 811 turbofans.
Performance: Max. speed, 820 mph (1 320 km/h) or Mach 1·1 at sea level, 1,057 mph (1 700 km/h) or Mach 1·6 at 32,810 ft (10 000 m); combat radius (with external fuel), 282 mls (454 km) LO-LO-LO, 440 mls (708 km) HI-LO-HI; unrefuelled ferry range, 2,190 mls (3 524 km).
Weights: Typical empty, 15,432 lb (7 000 kg); normal loaded (clean aircraft), 24,000 lb (11 000 kg); max. take-off, 34,000 lb (15 422 kg).
Armament: Two 30-mm Aden cannon and up to 10,000 lb (4 536 kg) of ordnance on five external stations. Provision for two Matra Magic AAMs on overwing stations or AIM-9P Sidewinder AAMs on underwing stations.
Status: The Jaguar International was developed jointly by British Aerospace in the UK and Dassault-Breguet in France. The first of eight Jaguar prototypes was flown on 8 September 1968, and 202 (including 37 two-seaters) of the basic version were delivered to the RAF and 200 (including 40 two-seaters) to the *Armée de l'Air*.
Notes: Although manufactured jointly with Dassault-Breguet, the Jaguar International is assembled by British Aerospace which has been responsible for supplying 12 each to Ecuador and Oman, and 40 (including five two-seaters) to India, the last of those for the last-mentioned country having been delivered in November 1982. Oman is to receive a further 12 during 1983, and 45 are being assembled by HAL in India, the first of these having flown on 31 March 1982. Current planning calls for HAL to produce a further 31 aircraft primarily from components of indigenous manufacture.

SEPECAT JAGUAR INTERNATIONAL

Dimensions: Span, 28 ft 6 in (8,69 m); length, 50 ft 11 in (15,52 m); height, 16 ft 0½ in (4,89 m); wing area, 280·3 sq ft (24,18 m²).

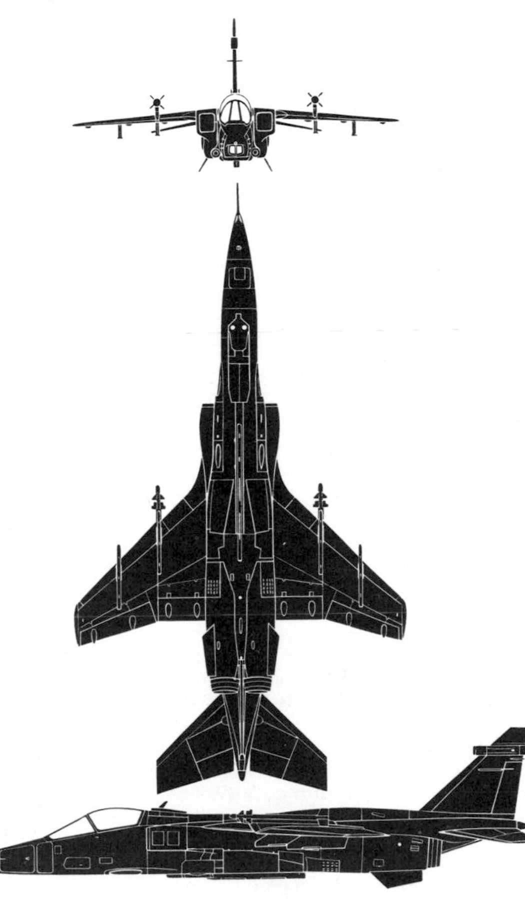

SHANGHAI Y-10

Country of Origin: China.
Type: Medium-haul commercial transport.
Power Plant: Four 19,000 lb st (8 618 kgp) Pratt & Whitney JT3D-7 turbofans.
Performance: Max. cruising speed, 605 mph (974 km/h) at 24,930 ft (7 600 m); long-range cruise, 570 mph (917 km/h); initial climb, 3,937 ft/min (20 m/sec); service ceiling, 40,450 ft (12 330 m); range (with 178 passengers), 3,455 mls (5 560 km).
Weights: Max. take-off, 224,867 lb (102 000 kg).
Accommodation: Flight crew of four and 178 passengers six-abreast with central aisle, or 124 passengers in mixed layout.
Status: The first of two Y-10 prototypes was flown on 26 September 1980, this following protracted testing of a structural test specimen.
Notes: Bearing a close resemblance to the Boeing 707, but marginally smaller, the Y-10 is the first jet transport of Chinese design and the engines by which it is powered were, in fact, acquired as spares for those powering the Boeing 707s operated by CAAC. Despite its close resemblance to the Boeing design, the Y-10 is not a copy, but no plans for series production have yet been announced. However, it would seem improbable that the Y-10 has been developed, as has been suggested, solely to demonstrate the capability of the Chinese aircraft industry to create an aircraft in this category. Reports of Chinese interest in acquisition of the CFM56 engine have been linked with proposals to utilise this power plant in a series version of the Y-10. During the initial flight test programme, the first Y-10 flew, on 8 December 1981, from Shanghai to Beijing (Peking) at an average speed of 528 mph (850 km/h).

SHANGHAI Y-10

Dimensions: Span, 138 ft 7 in (42,24 m); length, 140 ft 10 in (42,93 m); height, 44 ft 0⅓ in (13,42 m); wing area, 2,632·94 sq ft (244,60 m²).

SHORTS 330-200/SHERPA

Country of Origin: United Kingdom.
Type: Regional airliner and commercial freighter.
Power Plant: Two 1,198 shp Pratt & Whitney (Canada) PT6A-45R turboprops.
Performance: High-speed cruise, 232 mph (373 km/h) at 10,000 ft (3 050 m); long-range cruise, 184 mph (296 km/h); range (with 30 passengers and reserves), 320 mls (515 km), (with max. fuel and reserves), 775 mls (1 247 km); initial climb rate, 1,180 ft/min (5,99 m/sec).
Weights: Operational empty, 15,000 lb (6 805 kg); max. take-off, 22,900 lb (10 387 kg).
Accommodation: Flight crew of two and standard arrangement for 30 passengers in 10 rows three-abreast, or (Sherpa) up to three LD3 standard air freight containers.
Status: Engineering prototype of Shorts 330 flown on 22 August 1974, with production prototype following on 8 July 1975. First production aircraft flown on 15 December 1975, with customer deliveries commencing mid 1976. At the beginning of 1983, 88 had been delivered against orders and options for 112. A commercial freighter version, the Sherpa, entered flight test on 23 December 1982.
Notes: The 330-200 offers a number of product improvements over the basic 330, and the Sherpa freighter is based on the 330-200 airframe, having similar PT6A-45R engines but introducing a full-width rear loading door and a roller conveyor system. A proposed military equivalent, the UTT (Utility Tactical Transport) will not offer the rear loading door of the Sherpa and will be adapted to provide for a variety of military roles.

SHORTS 330-200/SHERPA

Dimensions: Span, 74 ft 8 in (22,76 m); length, 58 ft $0\frac{1}{2}$ in (17,69 m); height, 16 ft 3 in (4,95 m); wing area, 453 sq ft (42,10 m²).

SHORTS 360

Country of Origin: United Kingdom.
Type: Regional airliner.
Power Plant: Two 1,327 shp Pratt & Whitney (Canada) PT6A-65R turboprops.
Performance: High-speed cruise, 243 mph (391 km/h) at 10,000 ft (3 050 m); range at max. cruise (with max. payload and allowances for 100-mile/160-km diversion and 45 min hold), 360 mls (579 km), (with max. fuel and same reserves), 860 mls (1 384 km).
Weights: Operational empty, 16,600 lb (7 530 kg); max. take-off, 25,700 lb (11 657 kg).
Accommodation: Flight crew of two with standard arrangement for 36 passengers in 11 rows three-abreast. Baggage compartments in nose and aft of cabin.
Status: The prototype Shorts 360 flew for the first time on 1 June 1981, the first production aircraft following on 19 August 1982. First customer delivery (to Allegheny) on 11 November 1982, and four aircraft delivered by beginning of 1983, against 28 orders and options. Combined production rate of Shorts 330 and 360 scheduled to increase from three to five monthly during 1983.
Notes: The Shorts 360 is a growth version of the 330 (see pages 184–5) and differs from its progenitor primarily in having a 3-ft (91-cm) cabin stretch ahead of the wing and an entirely redesigned rear fuselage and tail assembly. The fuselage lengthening permits the insertion of two additional rows of three seats in the main cabin, and the lower aerodynamic drag by comparison with the earlier aircraft contributes to a higher performance.

SHORTS 360

Dimensions: Span, 74 ft 10 in (22,81 m); length, 70ft 10in (21,59m); height, 23ft 8in (7,21 m); wing area, 454 sq ft (42,18 m²).

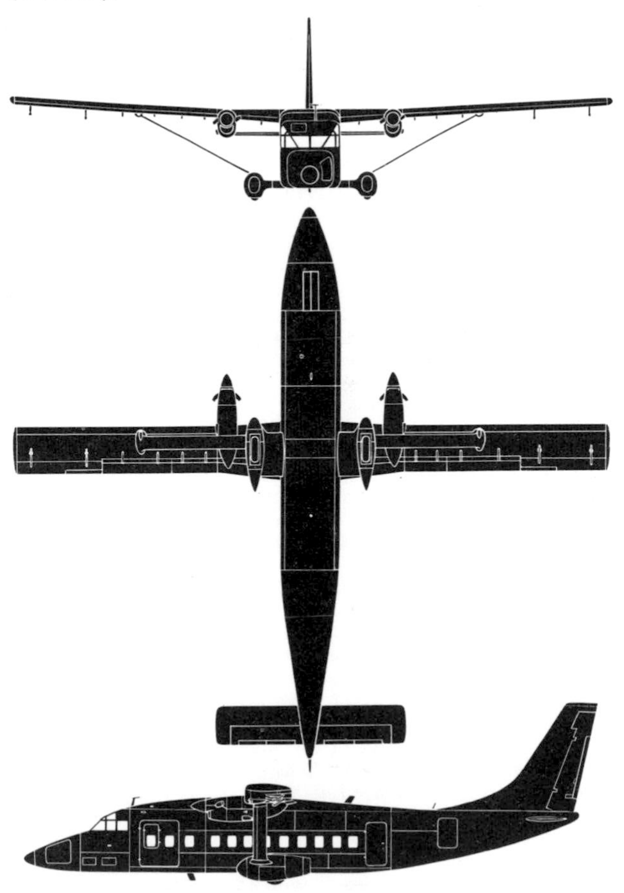

SIAI MARCHETTI S.211

Country of Origin: Italy.
Type: Tandem two-seat basic trainer.
Power Plant: One 2,500 lb st (1 134 kgp) Pratt & Whitney (Canada) JT15D-4C turbofan.
Performance: Max. speed, 449 mph (723 km/h) at 25,000 ft (7 620 m); max. cruise, 437 mph (704 km/h) at 25,000 ft (7 620 m); max. initial climb, 4,950 ft/min (25,15 m/sec); service ceiling, 42,000 ft (12 800 m); max. range (internal fuel and 30 min reserves), 1,187 mls (1 910 km) at 30,000 ft (9 145 m), (with two external tanks), 1,673 mls (2 693 km).
Weights: Empty, 3,186 lb (1 445 kg); normal loaded, 5,070 lb (2 300 kg); max. take-off, 6,173 lb (2 800 kg).
Armament: (Weapons training and light strike) Four wing stations each stressed for loads up to 660 lb (300 kg) inboard and 330 lb (150 kg) outboard, with maximum external load of 1,320 lb (600 kg).
Status: First of three prototypes flown on 10 April 1981. No firm orders had been announced by the beginning of 1983, although options had been taken on some 30 aircraft by the air arms of three countries (reportedly including Singapore) and the first production aircraft is scheduled for completion during the second quarter of 1983. A production rate of five per month is expected to be attained during 1984.
Notes: The S.211 has been developed by the Siai Marchetti component of the Agusta group as a private venture and as an attempt to arrest the upward spiralling cost of training a military pilot. The S.211 has less than half the empty weight of other new-generation jet trainers and is barely heavier than such turboprop trainers as the Pilatus PC-7. It is nevertheless a comparatively sophisticated design with a supercritical wing.

SIAI MARCHETTI S.211

Dimensions: Span, 26 ft $2\frac{7}{8}$ in (8,00 m); length, 30 ft $5\frac{1}{3}$ in (9,28 m); height, 12 ft $2\frac{3}{4}$ in (3,73 m); wing area, 135·63 sq ft (12,60 m²).

SIAI MARCHETTI SF.260TP

Country of Origin: Italy.
Type: Side-by-side two-seat primary/basic trainer.
Power Plant: One 350 shp Allison 250B-17C turboprop.
Performance: (At 2,645 lb/1 200 kg) Max. speed, 237 mph (382 km/h) at sea level; cruise (75% power), 230 mph (371 km/h) at 10,000 ft (3 050 m), 219 mph (352 km/h) at 20,000 ft (6 095 m); initial climb, 2,170 ft/min (11 m/sec); service ceiling, 28,000 ft (8 535 m); max. range (with 30 min reserves), 590 mls (950 km).
Weights: Empty equipped, 1,753 lb (795 kg); max. take-off, 2,645 lb (1 200 kg).
Armament: (SF.260TP Warrior for armament training and light strike) Four wing ordnance stations for max. of 661 lb (300 kg) including 7,62-mm gun pods, bombs of up to 331 lb (150 kg) and pods for rockets of up to 68-mm calibre.
Status: Prototype SF.260TP flown July 1980, with deliveries commencing 1982, initial customers including Ghana Air Force (eight plus option on further dozen) and the Dubai Air Wing (six), plus six for civil training in Haiti.
Notes: A turboprop-powered derivative of the highly successful piston-engined SF.260, of which more than 450 have been supplied to military customers and in excess of 100 to civil customers, having been in continuous production since 1966, the SF.260TP adheres closely structurally to its predecessor, comparatively few airframe modifications having taken place aft of the firewall. Developed primarily to overcome the difficulty of obtaining Avgas in certain areas of the world, the SF.260TP is being offered, like its piston-engined predecessor, in armed form as the Warrior for weapons training and light strike tasks. Production of the basic SF.260 with the Avco Lycoming O-540-E4A5 piston engine is continuing.

SIAI MARCHETTI SF.260TP

Dimensions: Span, 27 ft $4\frac{3}{4}$ in (8,35 m); length, 24 ft $3\frac{1}{2}$ in (7,40 m); height, 7 ft $10\frac{7}{8}$ in (2,41 m); wing area, 108·72 sq ft (10,10 m²).

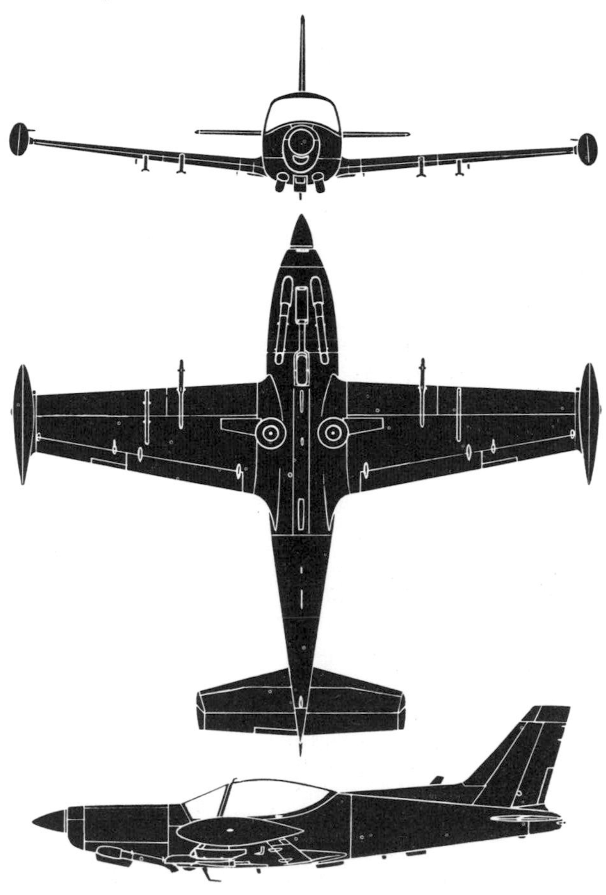

SLINGSBY T67M FIREFLY 160M

Country of Origin: United Kingdom.
Type: Side-by-side two-seat primary trainer.
Power Plant: One 160 hp Avco Lycoming AEIO-320-D1B four-cylinder horizontally-opposed engine.
Performance: (At aerobatic weight) Max. speed, 152 mph (245 km/h) at sea level; cruise, 136 mph (219 km/h) at 8,000 ft (2 440 m); initial climb, 1,150 ft/min (5,8 m/sec); endurance (at 65% power with reserves), 2·75 hrs.
Weights: Typical empty, 1,350 lb (612 kg); max. aerobatic, 1,800 lb (817 kg); max. take-off, 1,900 lb (862 kg).
Status: The first T67M Firefly entered flight test on 5 December 1982, with certification scheduled for March 1983. Production deliveries scheduled to commence during the course of 1983, an initial production order for 10 examples having been placed by Specialist Flying Training to which first series aircraft will be delivered. Current planning call for production to build up to 50 per year.
Notes: The T67M Firefly is an all-British, all-plastics and more powerful derivative of the French Fournier RF-6B, 10 examples of this wooden 120 hp two-seater having been built by Slingsby under licence as the T67A. One of the T67As was re-engined with the 160 hp unit and constant-speed propeller of the T67M to initiate certification trials on 15 July 1982. The T67M has a structure built entirely of glass-reinforced plastics (GRP), and planned derivatives include the T68 which will mate the GRP wing of the T67M with a new four-seat fuselage and a 200 hp engine. Consideration is being given to a version of the T67M with a fully-retractable undercarriage.

SLINGSBY T67M FIREFLY 160M

Dimensions: Span, 34 ft 9 in (10,60 m); length, 23 ft 0 in (7,10 m); height, 8 ft 3 in (2,51 m); wing area, 136 sq ft (12,63 m²).

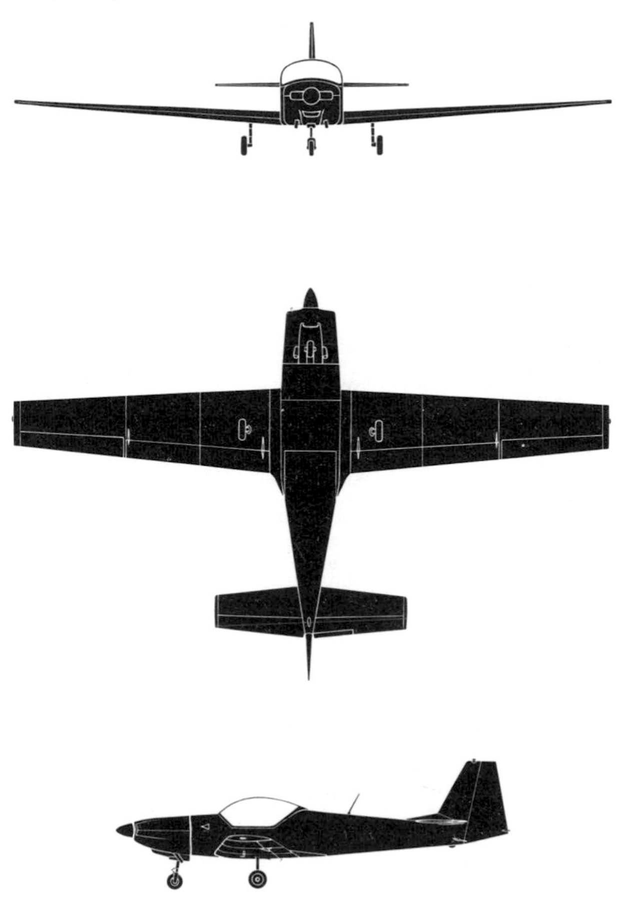

SOCATA TB 20 TRINIDAD

Country of Origin: France.
Type: Light cabin monoplane.
Power Plant: One 250 hp Avco Lycoming IO-540-C4D5D six-cylinder horizontally-opposed engine.
Performance: Max. speed, 193 mph (312 km/h); max. cruise (75% power), 188 mph (303 km/h) at 8,500 ft (2 600 m); econ cruise (65% power), 184 mph (296 km/h) at 12,000 ft (3 660 m); initial climb, 1,260 ft/min (6,4 m/sec); range (45 min reserves), 1,019 mls (1 640 km) at 75% power, 1,109 mls (1 785 km) at 65% power.
Weights: Empty, 1,710 lb (776 kg); max. take-off, 2,937 lb (1 332 kg).
Accommodation: Pilot and passenger with dual controls side by side in individual seats and two individual seats or bench-type seat for three at rear of cabin.
Status: Prototype TB 20 flown on 14 November 1980, with certification obtained early 1982, with production deliveries commencing in spring of that year.
Notes: The TB 20 is the latest in a new series of all-metal light aircraft initiated in 1977 with the TB 10 Tobago (see 1980 edition), a 180 hp fixed-undercarriage aircraft which continues in production alongside the more powerful TB 20 which introduces a retractable undercarriage. Also in production is a four-seat 160 hp version of the TB 10 known as the TB 9 Tampico. All three models are essentially similar structurally and possess similar external dimensions apart from slightly enlarged horizontal tail surfaces which have been applied to the TB 20 which also features an increase in wing fuel capacity. All TB series aircraft adopted new wingtip fairings from 1982.

SOCATA TB 20 TRINIDAD

Dimensions: Span, 32 ft $1\frac{3}{4}$ in (9,80 m); length, 25 ft $3\frac{1}{8}$ in (7,70 m); height, 9 ft 6 in (2,90 m); wing area, 128·1 sq ft (11,90 m²).

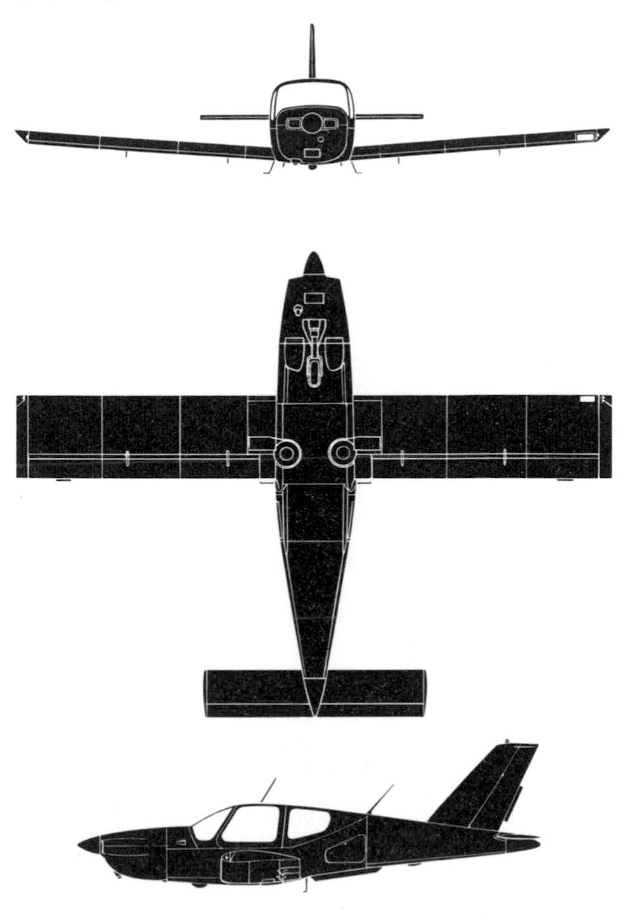

SOKO (CNIAR IAR-93A) ORAO

Countries of Origin: Yugoslavia and Romania.
Type: Single-seat close air support fighter.
Power Plant: Two 4,000 lb st (1 814 kgp) Turbomecanica-built Rolls-Royce Viper Mk 632-41 turbojets.
Performance: Max. speed, 700 mph (1 126 km/h) or Mach 0·92 at sea level, 665 mph (1 070 km/h) or Mach 0·95 at 25,000 ft (7 620 m); initial climb, 14,880 ft/min (76 m/sec); service ceiling, 42,650 ft (13 000 m); combat radius (with 4,410-lb/2 000-kg ordnance), 224 mls (360 km) HI-LO-HI, 186 mls (300 km) LO-LO-LO.
Weights: Empty equipped, 12,570 lb (5 700 kg); clean loaded, 18,960 lb (8 600 kg); max. take-off, 23,130 lb (10 500 kg).
Armament: Two 23-mm cannon and up to 5,510 lb (2 500 kg) of bombs, missile pods and air-to-surface missiles on one fuselage centreline and four wing stations.
Status: Two prototypes flown 31 October 1974 (one in Yugoslavia and one in Romania), followed by two two-seat operational trainer prototypes on 29 January 1977, and pre-production series of 40 (with non-afterburning engines) during 1981 from two assembly lines (in Yugoslavia and Romania). Deliveries of series model (with afterburning Viper Mk 633s) to both Yugoslav and Romanian air forces anticipated 1984–85.
Notes: The Orao (Eagle) has been developed as a joint programme by SOKO in Yugoslavia and CNIAR in Romania, with final assembly lines at Mostar and Craiova. Manufacture of the airframe and systems is shared between the two countries with no component duplication, the Viper turbojets being licence-manufactured in Bucharest. The two-seat conversion training version incorporates a second cockpit within a lengthened fuselage. Total of 165 of series version (IAR-93B) with afterburning scheduled to follow 20 IAR-93As into Romanian service.

SOKO (CNIAR IAR-93) ORAO

Dimensions: Span, 31 ft 7 in (9,63 m); length, 42 ft 10 in (14,88 m); height, 14 ft 7¼ in (4,45 m); wing area, 279·86 sq ft (26,00 m²).

SOKO SUPER GALEB

Country of Origin: Yugoslavia.
Type: Tandem two-seat basic/advanced trainer and light strike aircraft.
Power Plant: One 4,000 lb st (1 814 kgp) Turbomecanica-built Rolls-Royce Viper Mk 632-41 turbojet.
Performance: (Estimated) Max. speed, 550 mph (885 km/h) at sea level, 500 mph (805 km/h) at 30,000 ft (9 145 m); initial climb, 7,000 ft/min (35,5 m/sec); service ceiling, 45,000 ft (13 715 m).
Weights: Approx. normal loaded weight, 10,000 lb (4 536 kg).
Armament: (Weapons training and light strike) Four wing ordnance stations for cannon or rocket pods, surface-to-air missiles or bombs up to 500 lb (226,8 kg).
Status: The Super Galeb reportedly entered flight test during the course of 1981, several prototypes and pre-series examples being under test at the beginning of 1983.
Notes: The Super Galeb has been developed as a potential successor to the G2A Galeb (Seagull) serving with the Yugoslav Air Force and is closely comparable in size, weight and performance with the Italian Aermacchi MB-339A. The Viper Mk 632-41 turbojets of the Super Galeb are manufactured under license in Rumania by Turbomecanica. Possessing no design commonality with the Galeb, the Super Galeb bears some resemblance to the British Aerospace Hawk, and it may be assumed that, as with the earlier trainer, a single-seat dedicated light attack version will be evolved. The data provided on these pages should be considered provisional as little information relating to the Super Galeb had been officially revealed at the time of closing for press.

SOKO SUPER GALEB

Dimensions: (Estimated) Span, 29 ft 6 in (9,00 m); length, 38 ft 0 in (11,60 m); height, 13 ft 0 in (4,00 m).

SUKHOI SU-17 FITTER

Country of Origin: USSR.
Type: Single-seat (Fitter-C, D and H) attack and counterair aircraft, and (Fitter-F and J) multi-role fighter.
Power Plant: (Fitter-C, D and H) One 17,195 lb st (7 800 kgp) dry and 24,700 lb st (11 200 kgp) reheat Lyulka AL-21F turbojet, or (Fitter-F and J) 17,635 lb st (8 000 kgp) dry and 25,350 lb st (11 500 kgp) Tumansky R-29B turbofan.
Performance: (Estimated for Fitter-H) Max. speed (short-period dash), 1,430 mph (2 300 km/h) or Mach 2·17 at 39,370 ft (12 000 m), (sustained), 808 mph (1 300 km/h) or Mach 1·06 at sea level, 1,190 mph (1 915 km/h) or Mach 1·8 at 39,370 ft (12 000 m); combat radius (drop tanks on outboard wing pylons and 4,410 lb/2 000 kg of external ordnance), 320 mls (515 km) LO-LO-LO, 530 mls (853 km) HI-LO-HI.
Weights: (Fitter-H) Max. take-off, 39,022 lb (17 700 kg).
Armament: Two 30-mm NR-30 cannon and max. external ordnance load of 7,716 lb (3 500 kg).
Status: Variable-geometry derivative of fixed-geometry Su-7 (Fitter-A) first flown as technology demonstrator in 1966 as S-22I (Fitter-B). Initial series as Su-17 (Fitter-C) entered Soviet service in 1971, and subsequently supplied as the Su-20 to Algeria, Egypt, Iraq, Poland and Vietnam. Upgraded model with lengthened forward fuselage (Fitter-D) followed into Soviet service in 1976, and re-engined (R-29B) export version (Fitter-F) exported to Peru as Su-22. Extensively revised version (Fitter-H) introduced 1979, and re-engined (R-29B) export equivalent supplied to Libya (Fitter-J) as Su-22.
Notes: The Lyulka-powered Fitter-H (opposite page) and Tumansky-powered Fitter-J (illustrated above) are current production versions of the Su-17, export versions of the Su-17 with the respective engines are referred to as Su-20 and Su-22.

SUKHOI SU-17 FITTER

Dimensions: (Estimated) Span (28 deg sweep), 45 ft 0 in (13,70 m), (68 deg sweep), 32 ft 6 in (9,90 m); length (including probe) 58 ft 3 in (17,75 m); height, 15 ft 5 in (4,70 m); wing area, 410 sq ft (38,00 m²).

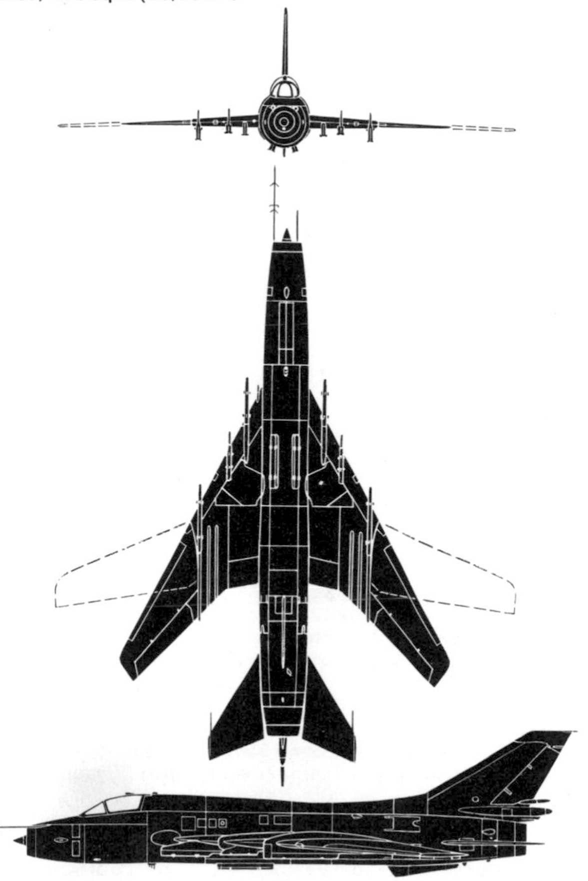

SUKHOI SU-24 (FENCER)

Country of Origin: USSR.
Type: Deep penetration interdictor and strike aircraft.
Power Plant: Two 17,635 lb st (8 000 kgp) dry and 25,350 lb st (11 500 kgp) reheat Tumansky R-29 turbofans.
Performance: (Estimated) Max. speed (clean), 915 mph (1 470 km/h) or Mach 1·2 at sea level, 1,520 mph (2 446 km/h) or Mach 2·3 above 36,000 ft (11 000 m); tactical radius (combat tanks and 4,400 lb/2 000 kg of ordnance), 1,050 mls (1 690 km) HI-LO-HI, 345 mls (555 km) LO-LO-LO.
Weights: (Estimated) Empty equipped, 41,890 lb (19 000 kg); max. take-off, 87,080 lb (39 500 kg).
Armament: One 23-mm six-barrel rotary cannon and one 30-mm cannon, and (short-range interdiction) up to 22 220-lb (100-kg) or 551-lb (250-kg) bombs, or 16 1,102-lb (500-kg) bombs. Various alternative missile loads.
Status: Prototype believed flown 1970, with initial operational status achieved late 1974. Production of approx 10 monthly as beginning of 1983, when 600–700 in service.
Notes: The first Soviet aircraft designed from the outset for interdiction and counterair missions, the Su-24 carries pilot and weapon systems operator side by side.

SUKHOI SU-24 (FENCER)

Dimensions: (Estimated) Span (16 deg sweep), 56 ft 6 in (17,25 m), (68 deg sweep), 33 ft 9 in (10,30 m); length (excluding probe) 65 ft 6 in (20,00 m) height, 18 ft 0 in (5,50 m); wing area, 452 sq ft (42,00 m²).

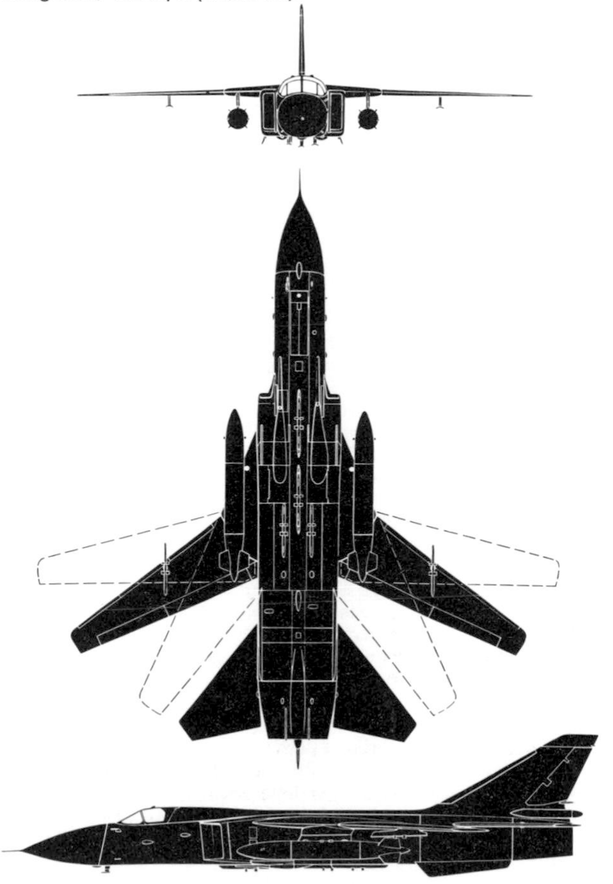

SUKHOI SU-25 (FROGFOOT)

Country of Origin: USSR.
Type: Single-seat close air support aircraft.
Power Plant: Two 11,240 lb st (5 100 kgp) Tumansky R-13-300 turbojets.
Performance: (Estimated) Max. speed, 420 mph (675 km/h) at sea level, 440 mph (708 km/h) at 10,000 ft (3 050 m); service ceiling, 35,000 ft (10 670 m); combat radius (with 10,000-lb/4 535-kg warload), 300 mls (480 km) at 25,000 ft (7 620 m).
Weights: (Estimated) Max. take-off, 38,000 lb (17 237 kg).
Armament: One six-barrel 23-mm rotary cannon. Ten external ordnance stations (eight wing and two fuselage) each capable of carrying a 1,102-lb (500-kg) bomb, a rocket pod or several types of anti-armour and anti-radiation missile.
Status: The Su-25 was first reported in 1980 and is believed to have entered flight test during 1977–78. Initial deliveries to the *Frontovaya Aviatsiya* are likely to have been made during 1980, and the aircraft was first deployed with an operational trials unit in Afghanistan during 1981.
Notes: The Su-25 is the Soviet equivalent of the Fairchild A-10A Thunderbolt II (see 1982 edition), and both single and two-seat versions are known to exist, the latter presumably being an operational training variant probably retaining combat capability. During operations with what is believed to be a trials unit in Afghanistan, the Su-25 has apparently been used to perfect the integration of low-level tactics of this fixed-wing aircraft with Mi-24 gunship helicopters. The Su-25 is expected to see large-scale deployment by *Frontovaya Aviatsiya* (Frontal Aviation) of the Soviet Air Forces during 1984–85.

SUKHOI SU-25 (FROGFOOT)

Dimensions: (Estimated) Span, 55 ft 0 in (16,75 m); length, 49 ft 0 in (14,95 m); height. 15 ft 0 in (4,57 m).

TRANSALL C.160NG

Countries of Origin: France and Federal Germany.
Type: Medium-range tactical transport.
Power Plant: Two 6,100 eshp Rolls-Royce/SNECMA Tyne RTy 20 Mk 22 turboprops.
Performance: Max. speed, 319 mph (513 km/h) at 16,000 ft (4 875 m); max. continuous cruise, 310 mph (499 km/h) at 20,000 ft (6 100 m); econ cruise, 282 mph (454 km/h); initial climb (at 108,355 lb/49 150 kg), 1,360 ft/min (6,9 m/sec); max. range (with max. payload and reserves of five per cent and 30 min), 1,150 mls (1 850 km), (with max. fuel and 17,637 lb/ 8 000 kg payload), 5,500 mls (8 854 km).
Weights: Operational empty, 61,728 lb (28 000 kg); max. take-off, 112,434 lb (51 000 kg).
Accommodation: Flight crew of three and up to 35,273 lb (16 000 kg) of freight, 66–88 paratroops, 93 fully-equipped troops, or up to 63 casualty stretchers and four medical attendants.
Status: First of second (relaunched) C.160 production series flown on 9 April 1981, and production was continuing at a rate of 0·7 monthly at the beginning of 1983 against orders for 25 for the *Armée de l'Air* and three for the Indonesian government, with 15 delivered, and eight to be produced during year. A further seven are to be procured by the *Armée de l'Air*.
Notes: The C.160NG is a relaunched production version of the original C.160, the first prototype of which was flown on 25 February 1963, with two further prototypes and 179 production aircraft having been built when manufacture terminated in 1972. The C.160NG embodies numerous refinements and improvements, 10 aircraft being equipped to serve as flight refuelling tankers and all having provision for in-flight refuelling. Manufacture is shared between France (Aérospatiale) and Federal Germany (MBB/VFW).

TRANSALL C.160NG

Dimensions: Span, 131 ft 3 in (40,00 m); length (excluding refuelling probe), 106 ft $3\frac{1}{2}$ in (32,40 m); height, 38 ft 2 in (11,65 m); wing area, 1,723·3 sq ft (160,10 m²).

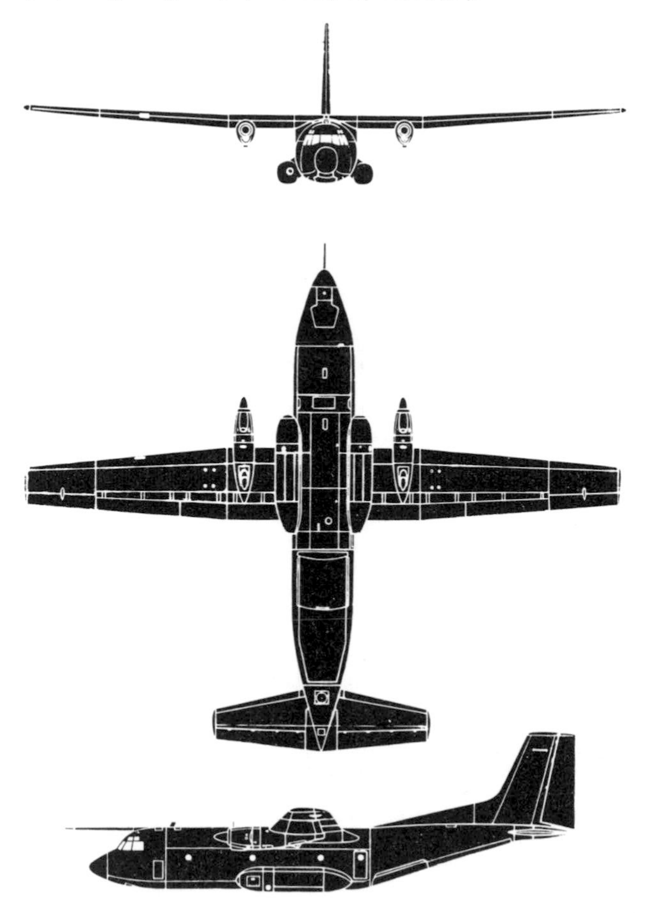

TUPOLEV (BACKFIRE-B)

Country of Origin: USSR.
Type: Medium-range strategic bomber and maritime strike/reconnaissance aircraft.
Power Plant: Two (estimated) 33,070 lb st (15 000 kgp) dry and 46,300 lb st (21 000 kgp) reheat Kuznetsov turbofans.
Performance: (Estimated) Max. speed (short-period dash), 1,265 mph (2 036 km/h) or Mach 1·91 at 39,370 ft (12 000 m), (sustained), 1,056 mph (1 700 km/h) or Mach 1·6 at 39,370 ft (12 000 m), 685 mph (1 100 km/h) or Mach 0·9 at sea level; combat radius (unrefuelled with single AS-4 ASM and high-altitude subsonic mission profile), 2,610 mls (4 200 km); max. unrefuelled combat range (with 12,345 lb/5 600 kg internal ordnance), 5,560 mls (8 950 km).
Weights: (Estimated) Max. take-off, 260,000 lb (118 000 kg).
Armament: Remotely-controlled tail barbette housing twin 23-mm NR-23 cannon. Internal load of free-falling weapons up to 12,345 lb (5 600 kg) or one AS-4 Kitchen inertially-guided stand-off missile housed semi-externally.
Status: Flight testing of initial prototype commenced late 1969, with pre-production series of up to 12 aircraft following in 1972–73. Initial version (Backfire-A) was built in small numbers only. Initial operational capability attained with Backfire-B in 1975–76, production rate of 30 annually being attained in 1977 and remaining constant at beginning of 1983, when 90–95 were in service with Soviet Long-range Aviation and a similar quantity with the Soviet Naval Air Force. An advanced version of Backfire with redesigned engine air intakes and presumably uprated engines has been reported under test, but its production status was uncertain at the beginning of 1983.

TUPOLEV (BACKFIRE-B)

Dimensions: (Estimated) Span (20 deg sweep), 115 ft 0 in (35,00 m), (55 deg sweep), 92 ft 0 in (28,00 m); length, 138 ft 0 in (42,00 m); height, 29 ft 6 in (9,00 m); wing area, 1,830 sq ft (170,00 m²).

VALMET PIK-23 TOWMASTER

Country of Origin: Finland.
Type: Side-by-side two-seat primary trainer and glider tug.
Power Plant: One 180 hp Avco Lycoming O-360-A4M four-cylinder horizontally-opposed engine.
Performance: Max. speed, 155 mph (250 km/h) at sea level; cruise (65% power), 137 mph (220 km/h) at 12,000 ft (3 660 m); initial climb rate, 1,142 ft/min (5,8 m/sec); range (65% power), 777 mls (1 250 km) at 12,000 ft (3 660 m).
Weights: Empty equipped, 1,301 lb (590 kg); max. take-off (normal cat), 1,918 lb (870 kg), (util cat), 1,750 lb (794 kg).
Status: The first of two prototypes of the PIK-23 was flown on 22 March 1982, and it is anticipated that series production will commence late 1983.
Notes: The latest in a series of light aircraft of all-composite construction designed by the Polytekuikkojen Ilmailukerho, the PIK-23 has been produced and is being marketed by Valmet, which, in addition to offering the complete aircraft, intends to offer kits of parts for club assembly. Developed from the PIK-19 Muhiru, which, first flown in March 1972, employed a similar composite structure (glass fibre-reinforced epoxy resin) which is claimed to offer ease of maintenance and repair in the field. While suitable for use as a civil or military *ab initio* trainer, the PIK-23 has been optimised for the glider-towing role and has a high power-to-weight ratio accordingly for good initial climb, even in hot and high conditions, when towing a glider. It also offers high descent rates and can perform a complete towing cycle every four minutes using a reeled towrope.

VALMET PIK-23 TOWMASTER

Dimensions: Span, 32 ft $9\frac{1}{2}$ in (10,00 m); length, 23 ft 7 in (7,19 m); height, 9 ft 6 in (2,90 m); wing area, 150·7 sq ft (14,00 m²).

YAKOVLEV YAK-36MP (FORGER-A)

Country of Origin: USSR.
Type: Shipboard VTOL air defence and strike fighter.
Power Plant: One (estimated) 17,640 lb st (8 000 kgp) lift/cruise turbojet plus two (estimated) 7,935 lb st (3 600 kgp) lift turbojets.
Performance: (Estimated) Max. speed, 648 mph (1 042 km/h) or Mach 0·85 at sea level, 595 mph (956 km/h) or Mach 0·9 above 36,000 ft (10 970 m); high-speed cruise, 560 mph (900 km/h) at 20,000 ft (6 095 m); tactical radius (internal fuel and 2,000 lb/900 kg ordnance), 230 mls (370 km) HI-LO-HI, 150 mls (240 km) LO-LO-LO, (recce mission with recce pod, two drop tanks and two AAMs), 340 mls (547 km).
Weights: (Estimated) Empty equipped, 16,500 lb (7 485 kg); max. take-off, 25,000 lb (11 340 kg).
Armament: Four underwing pylons with total (estimated) capacity of 2,205 lb (1 000 kg) for bombs, rockets, gun or rocket pods, or IR-homing AAMs.
Status: The Yak-36MP is believed to have flown in prototype form in 1971, and to have attained service evaluation status in 1976. At the beginning of 1983, the Yak-36MP was deployed aboard the carriers *Kiev*, *Minsk* and *Novorossiisk*.
Notes: The Yak-36 is unique among current service combat aircraft in that it possesses vertical take-off-and-landing capability but is incapable of performing rolling take-offs or landings. A tandem two-seat conversion trainer version (Forger-B) has an extended forward fuselage to accommodate a second cockpit, the nose being dropped to provide the two ejection seats with a measure of vertical stagger. Fences to prevent ingestion of reflected exhaust efflux by the lift jets were introduced 1981–82.

YAKOVLEV YAK-36MP (FORGER-A)

Dimensions: (Estimated) Span, 24 ft 7 in (7,50 m); length, 52 ft 6 in (16,00 m); height, 11 ft 0 in (3,35 m); wing area, 167 sq ft (15,50 m²).

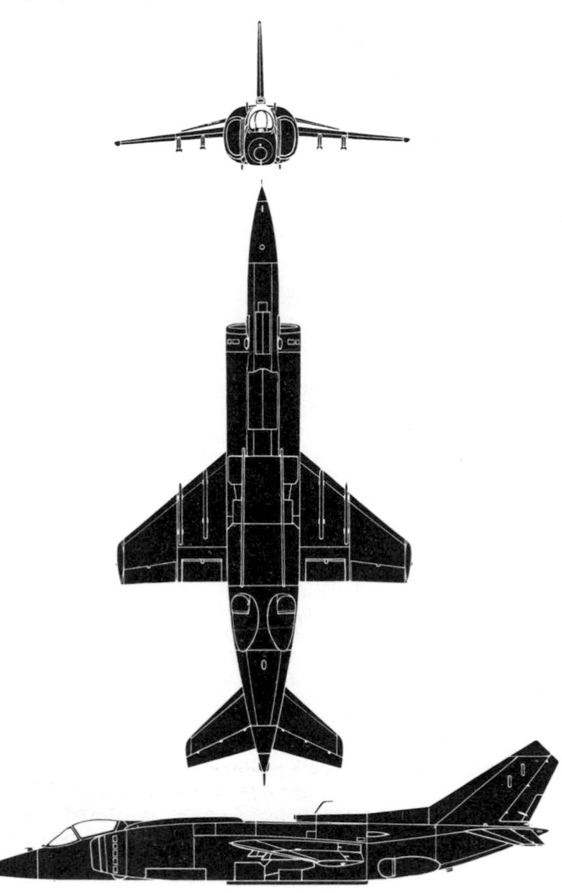

YAKOVLEV YAK-55

Country of Origin: USSR.
Type: Single-seat aerobatic competition aircraft.
Power Plant: One 360 hp Vedeneev M-14P nine-cylinder radial air-cooled engine.
Performance: Max. speed, 199 mph (320 km/h); cruising, 162 mph (260 km/h); initial climb, 3,150 ft/min (16 m/sec).
Weights: Empty, 1,411 lb (640 kg); max. take-off, 1,852 lb (840 kg).
Status: The Yak-55 has been developed by the "Progress" factory at Arseniev in the Soviet Far East, the specialist light aircraft development and manufacturing element of Aleksandr Yakovlev's bureau, and the third prototype made its international debut during the 1982 World Aerobatic Championships at Spiterberg, Austria.
Notes: Apart from its engine, the Yak-55 possesses no commonality with preceding Yakovlev "50" series light monoplanes, its principal innovatory features comprising a thick-section (18 per cent) mid-mounted wing, a blown cockpit canopy and a titanium spring leaf cantilever fixed undercarriage. Stressed for $+9g$ to $-9g$, the Yak-55 is a specialised competition aircraft and has been developed in parallel with the Yak-53, which, also introduced in 1982, is a progressive development of the Yak-50 (see 1977 edition) and Yak-52 (see 1979), respectively single- and two-seaters with tailwheel and nosewheel undercarriages. The Yak-53 is similar to the Yak-50 but utilises the nosewheel undercarriage of the Yak-52, and is to enter production during 1983 at both the Arseniev factory and at the IAv Bacau facility in Romania, the latter currently licence-manufacturing the Yak-52 under a Comecon programme. Development of the Yak-55 was continuing at the beginning of 1983.

YAKOVLEV YAK-55

Dimensions: Span, 26 ft $10\frac{1}{2}$ in (8,20 m); length, 24 ft $7\frac{1}{3}$ in (7,50 m); wing area, 153·93 sq ft (14,30 m²).

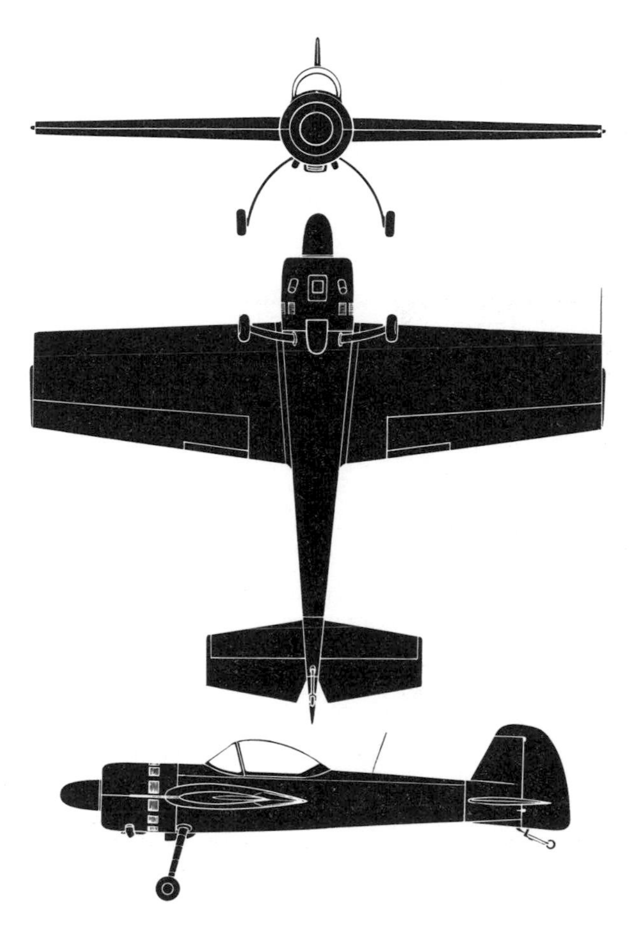

AÉROSPATIALE AS 332L SUPER PUMA

Country of Origin: France.
Type: Medium transport helicopter.
Power Plant: Two 1,755 shp Turboméca Makila turboshafts.
Performance: (At 18,080 lb/8 200 kg) Max. speed, 184 mph (296 km/h); max. cruise, 173 mph (278 km/h) at sea level, max. inclined climb. 1,810 ft/min (9,2 m/sec); hovering ceiling (in ground effect), 9,840 ft (3 000 m), (out of ground effect), 7,545 ft (2 300 m); range, 527 mls (850 km).
Weights: Empty, 9,635 lb (4 370 kg); normal loaded, 18,080 lb (8 200 kg); max. take-off, 19,840 lb (9 000 kg).
Dimensions: Rotor diam, 49 ft 5$\frac{3}{4}$ in (15,08 m); fuselage length, 48 ft 7$\frac{3}{4}$ in (14,82 m).
Notes: First flown on 10 October 1980, the AS 332L is a stretched (by 2·5 ft/76 cm) version of the basic Super Puma which is being produced in civil (AS 332C) and military (AS 332B) versions. The AS 332L and M are respectively civil and military variants of the stretched model, and the AS 332F is a navalised ASW version with an overall length of 42 ft 1$\frac{1}{3}$in (12,83 m) with rotor blades folded. Deliveries of the AS 332C began in October 1981 with the AS 332L following in December. Twenty Super Pumas (all versions) delivered by 30 June 1982, when production was rising to four monthly, this rate being scheduled to increase to five–six, with some 40 delivered by beginning of 1983 when more than 120 had been ordered. The AS 332B and C carry 20 troops and 17 passengers respectively.

AÉROSPATIALE SA 342 GAZELLE

Country of Origin: France.
Type: Five-seat light utility helicopter.
Power Plant: One 858 shp Turboméca Astazou XIVH turbo-shaft.
Performance: Max. speed, 192 mph (310 km/h); max. continuous cruise, 163 mph (263 km/h) at sea level; max. inclined climb, 1,675 ft/min (8,5 m/sec); hovering ceiling (in ground effect), 11,970 ft (3 650 m), (out of ground effect), 9,430 ft (2 875 m); range, 469 mls (754 km) at sea level.
Weights: Empty equipped, 2,149 lb (975 kg); max. take-off (normal), 4,190 lb (1 900 kg).
Dimensions: Rotor diam, 34 ft 5$\frac{1}{2}$ in (10,50 m); fuselage length (tail rotor included), 31 ft 2$\frac{3}{4}$ in (9,53 m).
Notes: A more powerful derivative of the SA 341 (592 shp Astazou IIIA), the SA 342 has been exported to Kuwait, Iraq, Libya and other Middle Eastern countries, and may be fitted with four or six HOT missiles, a 20-mm cannon and other weapons for the anti-armour role. One hundred and ten of the French Army's (166) SA 341F Gazelles are being equipped for HOT missiles as AS 341 Ms, and 128 SA 342Ms were in process of delivery to the French Army during 1982. Versions of the lower-powered SA 341 comprise the SA 341B (British Army), SA 341C (British Navy), SA 341D (RAF), SA 341G (civil) and SA 341H (military export). Orders for the SA 341 and 342 Gazelles totalled 1,072 by the beginning of 1983, and was continuing in collaboration with Westland.

AÉROSPATIALE AS 350 ECUREUIL

Country of Origin: France.
Type: Six-seat light general-purpose utility helicopter.
Power Plant: (AS 350B) One 641 shp Turboméca Arriel, or (AS 350D) 615 shp Avco Lycoming LTS 101-600A2 turboshaft.
Performance: (AS 350B) Max. speed, 169 mph (272 km/h) at sea level; cruise, 144 mph (232 km/h); max. inclined climb, 1,555 ft/min (7,9 m/sec); hovering ceiling (in ground effect), 9,678 ft (2 950 m), (out of ground effect), 7,382 ft (2 250 m); range, 435 mls (700 km) at sea level.
Weights: Empty, 2,348 lb (1 065 kg); max. take-off, 4,630 lb (2 100 kg).
Dimensions: Rotor diam, 35 ft 0¾ in (10,69 m); fuselage length (tail rotor included), 35 ft 9½ in (10,91 m).
Notes: The first Ecureuil (Squirrel) was flown on 27 June 1974 (with an LTS 101 turboshaft) and the second on 14 February 1975 (with an Arriel). The LTS 101-powered version (AS 350D) is being marketed in the USA as the AStar, some 260 having been delivered to US customers by the beginning of 1983, when production rate of both versions was running at 15–20 monthly with more than 600 delivered. The standard Ecureuil is a six-seater and features include composite rotor blades, a so-called Starflex rotor head, simplified dynamic machinery and modular assemblies to simplify changes in the field. The AS 350D AStar version is assembled and finished by Aérospatiale Helicopter at Grand Prairie, Alberta.

AÉROSPATIALE AS 355F ECUREUIL 2

Country of Origin: France.
Type: Six-seat light general-purpose utility helicopter.
Power Plant: Two 420 shp Allison 250-C20F turboshafts.
Performance: Max. speed, 169 mph (272 km/h) at sea level; max. cruise, 144 mph (232 km/h) at sea level; max. inclined climb, 1,614 ft/min (8,2 m/sec); hovering ceiling (out of ground effect), 7,900 ft (2 410 m); service ceiling, 14,800 ft (4 510 m); range, 470 mls (756 km) at sea level.
Weights: Empty, 2,778 lb (1 260 kg); max. take-off, 5,292 lb (2 400 kg).
Dimensions: Rotor diam, 35 ft 0¾ in (10,69 m); fuselage length (tail rotor included), 35 ft 9½ in (10,91 m).
Notes: Flown for the first time on 27 September 1979, the Ecureuil 2 employs an essentially similar airframe and similar dynamic components to those of the single-engined AS 350 Ecureuil (see page 218), and is intended primarily for the North American market on which it is known as the TwinStar. Deliveries of the Ecureuil 2/TwinStar commenced in July 1981, with more than 220 delivered by the beginning of 1983. From the first quarter of 1982, the production model has been the AS 355F which possesses a higher maximum take-off weight than the AS 355E that it has succeeded. The AS 355F has main rotor blades of increased chord, twin-body servo command units and two electrical generators. The AS 355E may be retrofitted to F standard. Total orders for the Ecureuil 2 worldwide exceeded 480 by the beginning of 1983; production was 20 monthly.

AÉROSPATIALE SA 365 DAUPHIN 2

Country of Origin: France.
Type: Multi-purpose and transport helicopter.
Power Plant: Two 700 shp Turboméca Arriel 1C turboshafts.
Performance: (SA 365N) Max. speed, 190 mph (305 km/h); max. continuous cruise, 173 mph (278 km/h) at sea level; max. inclined climb, 1,279 ft/min (6,5 m/sec); hovering ceiling (in ground effect), 3,296 ft (1 005 m), (out of ground effect), 3,116 ft (950 m); range, 548 mls (882 km) at sea level.
Weights: Empty, 4,288 lb (1 945 kg); max. take-off, 8,487 lb (3 850 kg).
Dimensions: Rotor diam, 39 ft 1½ in (11,93 m); fuselage length (including tail rotor), 37 ft 6⅓ in (11,44 m).
Notes: Flown as a prototype on 31 March 1979, the SA 365 is the latest derivative of the basic Dauphin (see 1982 edition), and is being manufactured in four versions, the 10–14-seat commercial SA 365N (illustrated above), the military SA 365M which can transport 13 commandos and carry eight HOT missiles, the navalised SA 365F with folding rotor, Agrion radar and four AS 15TT anti-ship missiles (20 ordered by Saudi Arabia for delivery from 1983) and the SA 366G, an Avco Lycoming LTS 101-750-powered search and rescue version for the US Coast Guard as the HH-65A Seaguard. Ninety of the last version are being procured by the US Coast Guard, with completion in 1985. Production of the SA 365N is scheduled to attain 10 monthly during 1983, and some 40 had been delivered by the beginning of that year.

AGUSTA A 109A MK II

Country of Origin: Italy.
Type: Eight-seat light utility helicopter.
Power Plant: Two 420 shp Allison 250-C20B turboshafts.
Performance: (At 5,402 lb/2 450 kg) Max. speed, 193 mph (311 km/h); max. continuous cruise, 173 mph (278 km/h); range cruise, 143 mph (231 km/h); max. inclined climb rate, 1,820 ft/min (9,25 m/sec); hovering ceiling (in ground effect), 9,800 ft (2 987 m), (out of ground effect), 6,800 ft (2 073 m); max. range, 356 mls (573 km).
Weights: Empty equipped, 3,125 lb (1 418 kg); max. take-off, 5,730 lb (2 600 kg).
Dimensions: Rotor diam, 36 ft 1 in (11,00 m); fuselage length, 35 ft 2½ in (10,73 m).
Notes: The A 109A Mk II is an improved model of the basic A 109A, the first of four prototypes of which flew on 4 August 1971, with customer deliveries commencing late 1976. Some 300 A 109As had been ordered by the beginning of 1983, at which time more than 170 had been delivered with production running at four–five monthly. The Mk II, which supplanted the initial model in production during 1981, has been the subject of numerous detail improvements, the transmission rating of the combined engines being increased from 692 to 740 shp, and the maximum continuous rating of each engine from 385 to 420 shp. An anti-armour version has been procured by Argentine, Libyan and Yugoslav forces. This can carry four or eight Hughes TOW anti-armour missiles.

AGUSTA A 129 MANGUSTA

Country of Origin: Italy.
Type: Two-seat light attack helicopter.
Power Plant: Two 815 shp Rolls-Royce Gem 2-2 Mk 1004D turboshafts.
Performance: (Estimated) Max. speed, 173 mph (278 km/h); cruise (TOW configuration at 8,377 lb/3 800 kg), 149 mph (240 km/h) at 5,740 ft (1 750 m); max. inclined climb (at 8,377 lb/3 800 kg), 1,900 ft/min (9,65 m/sec); hovering ceiling at 8,090 lb/3 670 kg), (in ground effect), 10,795 ft (3 290 m), (out of ground effect), 7,840 ft (2 390 m).
Weights: Mission, 8,080 lb (3 665 kg); max. take-off, 8,377 lb (3 800 kg).
Dimensions: Rotor diam, 39 ft $0\frac{1}{2}$ in (11,90 m); fuselage length, 40 ft $3\frac{1}{4}$ in (12,27 m).
Notes: The A 129 Mangusta (Mongoose) dedicated attack and anti-armour helicopter with full night/bad weather combat capability has been developed to an Italian Army requirement. The first of four flying prototypes is scheduled to commence flight test on 15 September 1983, and first deliveries are scheduled for mid-1986, 60 having been funded for the Italian Army with a further 30 on option. In typical anti-armour configuration, the A 129 will be armed with eight TOW missiles to which can be added 2·75-in (7-cm) rocket launchers for suppressive fire. A more advanced version, the A 129D, is being offered to Federal Germany to meet the PAH-2 requirement.

BELL MODEL 206B JETRANGER III

Country of Origin: USA.
Type: Five-seat light utility helicopter.
Power Plant: One 420 shp Allison 250-C20J turboshaft.
Performance: (At 3,200 lb/1 451 kg) Max. speed, 140 mph (225 km/h) at sea level; max. cruise, 133 mph (214 km/h) at sea level; hovering ceiling (in ground effect), 12,700 ft (3 871 m), (out of ground effect), 6,000 ft (1 829 m); max. range (no reserve), 360 mls (579 km).
Weights: Empty, 1,500 lb (680 kg); max. take-off, 3,200lb (1 451 kg).
Dimensions: Rotor diam, 33 ft 4 in (10,16 m); fuselage length, 31 ft 2 in (9,50 m).
Notes: Introduced in 1977, with deliveries commencing in July of that year, the JetRanger III differs from the JetRanger II which it supplants in having an uprated engine, an enlarged and improved tail rotor mast and more minor changes. Some 3,000 commercial JetRangers had been delivered by the beginning of 1983, both commercial and military versions (including production by licensees) totalling more than 7,500. A light observation version of the JetRanger for the US Army is designated OH-58 Kiowa and a training version for the US Navy is known as the TH-57A SeaRanger. The JetRanger is built by Agusta in Italy as the AB 206, and at the beginning of 1983, Agusta was producing the JetRanger at a rate of six monthly with approximately 1,000 delivered.

BELL MODEL 206L-3 LONGRANGER III

Country of Origin: USA.
Type: Seven-seat light utility helicopter.
Power Plant: One 650 shp Allison 250-C30P turboshaft.
Performance: (At 3,900 lb/1 769 kg) Max. speed, 144 mph (232 km/h); cruise, 136 mph (229 km/h) at sea level; hovering ceiling (in ground effect), 16,500 ft (5 030 m), (out of ground effect), 6,000 ft (1 830 m); range, 351 mls (565 km) at sea level.
Weights: Empty, 2,160 lb (980 kg); max. take-off, 4,150 lb (1 882 kg).
Dimensions: Rotor diam, 37 ft 0 in (11,28 m); fuselage length, 33 ft 3 in (10,13 m).
Notes: The Model 206L-3 LongRanger III is a stretched and more powerful version of the Model 206B JetRanger III, with longer fuselage, increased fuel capacity, an uprated engine and a larger rotor. The LongRanger is being manufactured in parallel with the JetRanger III and initial customer deliveries commenced in October 1975, prototype testing having been initiated on 11 September 1974. The LongRanger is available with emergency flotation gear and with a 2,000-lb (907-kg) capacity cargo hook. In the aeromedical or rescue role the Long Ranger can accommodate two casualty stretchers and two ambulatory casualties. The 206L-1 LongRanger II was introduced in 1978 and the uprated LongRanger III early in 1982, and production was 10 monthly at the beginning of 1983, with more than 800 LongRangers delivered.

BELL AH-1S HUEYCOBRA

Country of Origin: USA.
Type: Two-seat light attack helicopter.
Power Plant: One 1,800 shp Avco Lycoming T53-L-703 turboshaft.
Performance: Max. speed, 172 mph (277 km/h), (TOW configuration), 141 mph (227 km/h); max. inclined climb, 1,620 ft/min (8,23 m/sec); hovering ceiling TOW configuration (in ground effect), 12,200 ft (3 720 m); max. range, 357 mls (574 km).
Weights: (TOW configuration) Operational empty, 6,479 lb (2 939 kg); max. take-off, 10,000 lb (4 535 kg).
Dimensions: Rotor diam, 44 ft 0 in (13,41 m); fuselage length, 44 ft 7 in (13,59 m).
Notes: The AH-1S is a dedicated attack and anti-armour helicopter serving primarily with the US Army which had received 297 new-production AH-1S HueyCobras by mid-1981, plus 290 resulting from the conversion of earlier AH-1G and AH-1Q HueyCobras. Current planning calls for conversion of a further 372 AH-1Gs to AH-1S standards, and both conversion and new-production AH-1S HueyCobras are being progressively upgraded to "Modernised AH-1S" standard, the entire programme being scheduled for completion in 1985, resulting in a total of 959 "Modernised" AH-1S HueyCobras. In December 1979, one YAH-1S was flown with a four-bladed main rotor as the Model 249. The AH-1S is to be licence-built in Japan.

BELL AH-1T SEACOBRA

Country of Origin: USA.
Type: Two-seat light attack helicopter.
Power Plant: One 1,970 shp Pratt & Whitney T400-WV-402 coupled turboshaft.
Performance: (Attack configuration at 12,401 lb/5 625 kg) Max. speed, 181 mph (291 km/h) at sea level; average cruise, 168 mph (270 km/h); max. inclined climb, 2,190 ft/min (11,12 m/sec); hovering ceiling (out of ground effect), 5,350 ft (1 630 m); range, 276 mls (445 km).
Weights: Empty, 8,030 lb (3 642 kg); max. take-off, 14,000 lb (6 350 kg).
Dimensions: Rotor diam, 48 ft 0 in (14,63 m); fuselage length, 45 ft 3 in (13,79 m).
Notes: The SeaCobra is a twin-turboshaft version of the HueyCobra (see page 225), the initial model for the US Marine Corps having been the AH-1J (69 delivered of which two modified as AH-1T prototypes). The AH-1T features uprated components for significantly increased payload and performance, the first example having been delivered to the US Marine Corps on 15 October 1977, and a further 56 being delivered to the service of which 23 being modified to TOW configuration. Forty-four are to be procured by the USMC during Fiscal Years 1984 and 1985. The AH-1T has a three-barrel 20-mm cannon barbette under the nose, and four stores stations under the stub wings for seven- or 19-tube launchers, Minigun pods, etc.

BELL MODEL 214ST

Country of Origin: USA.
Type: Medium transport helicopter (19 seats).
Power Plant: Two 1,625 shp (limited to combined output of 2,250 shp) General Electric CT7-2A turboshafts.
Performance: Max. cruising speed, 164 mph (264 km/h) at sea level, 161 mph (259 km/h) at 4,000 ft (1 220 m); hovering ceiling (in ground effect), 12,600 ft (3 840 m), (out of ground effect), 3,300 ft (1 005 m); range (standard fuel), 460 mls (740 km).
Weights: Max. take-off (internal load), 15,500 lb (7 030 kg), (external jettisonable load), 16,500 lb (7 484 kg).
Dimensions: Rotor diam, 52 ft 0 in (15,85 m); fuselage length, 50 ft 0 in (15,24 m).
Notes: The Model 214ST (Super Transport) is a significantly improved derivative of the Model 214B BigLifter (see 1978 edition), production of which was phased out early 1981, and initial customer deliveries began early 1982. The Model 214ST test-bed was first flown in March 1977, and the first of three representative prototypes (one in military configuration and two for commercial certification) commenced its test programme in August 1979. Work on an initial series of 100 helicopters of this type commenced in 1981, with 16 delivered in 1982, 18–20 planned for delivery during 1983, and production tempo scheduled to reach three monthly in 1983. Alternative layouts are available for either 16 or 17 passengers.

BELL MODEL 222B

Country of Origin: USA.
Type: Eight/ten-seat light utility and transport helicopter.
Power Plant: Two 680 shp Avco Lycoming LTS 101-750C-1 turboshafts.
Performance: Max. cruising speed, 150 mph (241 km/h) at sea level, 146 mph (235 km/h) at 8,000 ft (2 400 m); max. climb, 1,730 ft/min (8,8 m/sec); hovering ceiling (in ground effect), 10,300 ft (3 135 m), (out of ground effect), 6,400 ft (1 940 m); range (no reserves), 450 mls (724 km) at 8,000 ft (2 400 m).
Weights: Empty equipped, 4,577 lb (2 076 kg); max. take-off (standard configuration), 8,250 lb (3 742 kg).
Dimensions: Rotor diam, 42 ft 0 in (12,80 m); fuselage length, 39 ft 9 in (12,12 m).
Notes: The first of five prototypes of the Model 222 was flown on 13 August 1976, an initial production series of 250 helicopters of this type being initiated in 1978, with production deliveries commencing in January 1980, and some 200 delivered by beginning of 1983, when production rate was two monthly. Several versions of the Model 222 are on offer or under development, these including an executive version with a flight crew of two and five or six passengers and the so-called "offshore" model with accommodation for eight passengers and a flight crew of two. Options include interchangeable skids. The Model 222B, deliveries of which commenced late 1982, has a larger main rotor and uprated power plant.

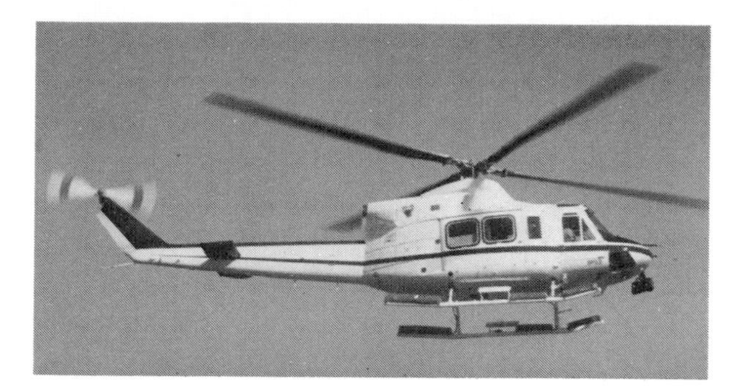

BELL MODEL 412

Country of Origin: USA.
Type: Fifteen-seat utility transport helicopter.
Power Plant: One 1,800 shp Pratt & Whitney PT6T-3B-1 turboshaft.
Performance: Max. speed, 149 mph (240 km/h) at sea level; cruise, 143 mph (230 km/h) at sea level, 146 mph (235 km/h) at 5,000 ft (1 525 m); hovering ceiling (in ground effect), 10,800 ft (3 290 m), (out of ground effect), 7,100 ft (2 165 m) at 10,500 lb/4 763 kg; max. range, 282 mls (454 km), (with auxiliary tanks), 518 mls (834 km).
Weights: Empty equipped, 6,070 lb (2 753 kg); max. take-off, 11,500 lb (5 216 kg).
Dimensions: Rotor diam, 46 ft 0 in (14,02 m); fuselage length, 41 ft 8½ in (12,70 m).
Notes: The Model 412, flown for the first time in August 1979, is an updated Model 212 (production of which was continuing at the beginning of 1983) with a new-design four-bladed rotor, a shorter rotor mast assembly, and uprated engine and transmission systems, giving more than twice the life of the Model 212 units. Composite rotor blades are used and the rotor head incorporates elastomeric bearings and dampers to simplify moving parts. An initial series of 200 helicopters was laid down with initial deliveries commencing February 1981 and 57 being delivered during the year. Licence manufacture is undertaken by Agusta in Italy, a military version being designated AB 412 Griffon.

BOEING VERTOL 234 CHINOOK

Country of Origin: USA.
Type: Commercial transport helicopter.
Power Plant: Two 4,075shp Avco Lycoming AL 5512 turboshafts.
Performance: Max. cruising speed (at 47,000 lb/21 318 kg), 167 mph (269 km/h) at 2,000 ft (610 m); range cruise, 155 mph (250 km/h); max. inclined climb, 1,350 ft/min (6,8m/sec); hovering ceiling (in ground effect), 9,150 ft (2 790 m), (out of ground effect), 4,900 ft (1 495 m); range (44 passengers and 45 min reserves), 627 mls (1 010 km), (max. fuel), 852 mls (1 371 km).
Weights: Empty, 24,449 lb (11 090 kg); max. take-off, 47,000 lb (21 318 kg).
Dimensions: Rotor diam (each), 60 ft 0 in (18,29 m); fuselage length, 52 ft 1 in (15,87 m).
Notes: Possessing an airframe based on the latest Model 414 military Chinook (see opposite), the Model 234 has been developed specifically for commercial purposes and two basic versions are offered, a long-range model described above and a utility model with fuel tank-housing side fairings removed. The first Model 234 was flown on 19 August 1980, certification being obtained mid-1981 and the first deliveries (to British Airways Helicopters) followed during the course of the year, primarily for North Sea oil rig support, six being delivered. Two will be delivered to Helikopter Service of Norway during 1983, when two will also be delivered to Arco Alaska.

BOEING VERTOL 414 CHINOOK

Country of Origin: USA.
Type: Medium transport helicopter.
Power Plant: Two 3,750 shp Avco Lycoming T55-L-712 turboshafts.
Performance: (At 45,400 lb/20 593 kg) Max. speed, 146 mph (235 km/h) at sea level; average cruise, 131 mph (211 km/h); max. inclined climb, 1,380 ft/min (7,0 m/sec); service ceiling, 8,400 ft (2 560 m); max. ferry range, 1,190 mls (1 915 km).
Weights: Empty, 22,591 lb (10 247 kg); max. take-off, 50,000 lb (22 680 kg).
Dimensions: Rotor diam (each), 60 ft 0 in (18,29 m); fuselage length, 51 ft 0 in (15,55 m).
Notes: The Model 414 as supplied to the RAF as the Chinook HC Mk 1 combines some features of the US Army's CH-47D (see 1980 edition) and features of the Canadian CH-147, but with provision for glassfibre/carbonfibre rotor blades. The first of 33 Chinook HC Mk 1s for the RAF was flown on 23 March 1980 and accepted on 2 December 1980, with deliveries continuing through 1981, three more being ordered in 1982 for 1984 delivery. The RAF version can accommodate 44 troops and has three external cargo hooks. During 1981, Boeing Vertol initiated the conversion to essentially similar CH-47D standards a total of 436 CH-47As, Bs and Cs. Licence manufacture of the Chinook is undertaken in Italy. Illustrated is one of three Model 414s delivered to Spain in 1982.

HUGHES 500MD DEFENDER II

Country of Origin: USA.
Type: Light gunship and multi-role helicopter.
Power Plant: One 420 shp Allison 250-C20B turboshaft.
Performance: (At 3,000 lb/1 362 kg) Max. speed, 175 mph (282 km/h) at sea level; cruise, 160 mph (257 km/h) at 4,000 ft (1 220 m); max. inclined climb, 1,920 ft/min (9,75 m/sec); hovering ceiling (in ground effect), 8,800 ft (2 682 m), (out of ground effect), 7,100 ft (2 164 m); max. range, 263 mls (423 km).
Weights: Empty, 1,295 lb (588 kg); max. take-off (internal load), 3,000 lb (1 362 kg), (with external load), 3,620 lb (1 642 kg).
Dimensions: Rotor diam, 26 ft 5 in (8,05 m); fuselage length, 21 ft 5 in (6,52 m).
Notes: The Defender II multi-mission version of the Model 500MD was introduced mid-1980 for 1982 delivery, and features a Martin Marietta rotor mast-top sight, a General Dynamics twin-Stinger air-to-air missile pod, an underfuselage 30-mm chain gun and a pilot's night vision sensor. The Defender II can be rapidly reconfigured for anti-armour target designation, anti-helicopter, suppressive fire and transport roles. The Model 500MD TOW Defender (carrying four tube-launched optically-tracked wire-guided anti-armour missiles) is currently in service with Israel (30), South Korea (25) and Kenya (15). Production of the 500 was 15 monthly at beginning of 1983.

HUGHES 500E

Country of Origin: USA.
Type: Five-seat light utility helicopter.
Power Plant: One 420 shp Allison 250-C20B turboshaft.
Performance: Max. cruising speed, 160 mph (258 km/h) at sea level, 155 mph (249 km/h) at 5,000 ft (1 525 m); max. inclined climb, 1,900 ft/min (9,65 m/sec); hovering ceiling (in ground effect), 8,500 ft (2 590 m), (out of ground effect), 7,500 ft (2 285 m); range, 330 mls (531 km) at 5,000 ft (1 525 m).
Weights: Empty, 1,320 lb (598 kg); max. take-off, 3,000 lb (1 360 kg).
Dimensions: Rotor diam, 26 ft 6 in (8,08 m); fuselage length, 23 ft 2½ in (7,07 m).
Notes: Evolved from the Model 500D and flown for the first time on 28 January 1982, the Model 500E and the more powerful Model 530E which commenced flight test in the following October, are characterised by a longer, recontoured nose which provides increased leg room for front seat occupants and 12 per cent increase in headroom for rear seat passengers, as well as various other refinements. The Model 530E differs in having a 650 shp Allison 250-C30 engine in place of the C20B and is intended for operation under hot-and-high conditions. The main rotor blades are 6 in (15 cm) longer. Customer deliveries of the Model 500E commenced in November 1982, and the Model 530E will follow from mid-1983.

HUGHES AH-64 APACHE

Country of Origin: USA.
Type: Tandem two-seat attack helicopter.
Power Plant: Two 1,690 shp General Electric T700-GE-701 turboshafts.
Performance: Max. speed, 191 mph (307 km/h); cruise, 179 mph (288 km/h); max. inclined climb, 3,200 ft/min (16,27 m/sec); hovering ceiling (in ground effect), 14,600 ft (4 453 m), (outside ground effect), 11,800 ft (3 600 m); service ceiling, 8,000 ft (2 400 m); max. range, 424 mls (682 km).
Weights: Empty, 9,900 lb (4 490 kg); primary mission, 13,600 lb (6 169 kg); max. take-off, 17,400 lb (7 892 kg).
Dimensions: Rotor diam, 48 ft 0 in (14,63 m); fuselage length, 48 ft 1⅞ in (14,70 m).
Notes: Winning contender in the US Army's AAH (Advanced Attack Helicopter) contest, the YAH-64 flew for the first time on 30 September 1975. Two prototypes were used for the initial trials, the first of three more with fully integrated weapons systems commenced trials on 31 October 1979, a further three following in 1980. Planned total procurement comprises 515 AH-64s through 1989, with a peak production rate of 12 monthly and deliveries commencing February 1984. The AH-64 is armed with a single-barrel 30-mm gun based on the chain-driven bolt system and suspended beneath the forward fuselage, and eight BGM-71A TOW anti-armour missiles may be carried, alternative amament including 16 Hellfire laser-seeking missiles.

KAMOV KA-25 (HORMONE A)

Country of Origin: USSR.
Type: Shipboard anti-submarine warfare helicopter.
Power Plant: Two 900 shp Glushenkov GTD-3 turboshafts.
Performance: (Estimated) Max. speed, 130 mph (209 km/h); normal cruise, 120 mph (193 km/h); max. range, 400 mls (644 km); service ceiling, 11,000 ft (3 353 m).
Weights: (Estimated) Empty, 10,500 lb (4 765 kg); max. take-off, 16,500 lb (7 484 kg).
Dimensions: Rotor diam (each), 51 ft 7½ in (15,74 m); approx. fuselage length, 35 ft 6 in (10,82 m).
Notes: Possessing a basically similar airframe to that of the Ka-25K (see 1973 edition) and employing a similar self-contained assembly comprising rotors, transmission, engines and auxiliaries, the Ka-25 serves with the Soviet Navy primarily in the ASW role but is also employed in the utility and transport roles. The ASW Ka-25 serves aboard the helicopter cruisers *Moskva* and *Leningrad*, and the carriers *Kiev* and *Minsk*, as well as with shore-based units. A search radar installation is mounted in a nose randome, but other sensor housings and antennae differ widely from helicopter to helicopter. There is no evidence that externally-mounted weapons may be carried. Each landing wheel is surrounded by an inflatable pontoon surmounted by inflation bottles. The Hormone-A is intended for ASW operations, the Hormone-B is used for over-the-horizon missile targeting, and the Hormone-C is a utility transport with operational equipment and weapons deleted.

KAMOV KA-32 (HELIX)

Country of Origin: USSR.
Type: Shipboard anti-submarine warfare helicopter.
Power Plant: Two (approx) 1,500-1,700 shp Glushenkov turboshafts.
Performance: (Estimated) Max. speed, 150 mph (241 km/h) at sea level; normal cruise, 130 mph (209 km/h); max. range, 600 mls (965 km).
Weights: (Estimated) Normal loaded, 20,000-21,000 lb (9 070-9 525 kg).
Dimensions: (Estimated) Rotor diam (each), 55 ft (16,75 m); fuselage length, 36 ft 4 in (12 00 m).
Notes: Retaining the pod-and-boom fuselage configuration and superimposed co-axial rotor arrangement of the Ka-25 Hormone (see page 235), the Ka-32 was first seen during Zapad-81 exercises held by WarPac forces in the Baltic in September 1981, and is believed to have flown in prototype form in 1979–80. Developed in both shipboard ASW and civil freight transportation versions simultaneously, the Ka-32 is substantially larger and more powerful than the preceding Kamov helicopter which it is presumably intended to supplant in Soviet Naval service, its greater internal capacity suggesting that an alternative mission to ASW may be that of assault troop transport for operation from *Berezina*-class replenishment ships. The Ka-32 would appear suitable as a replacement for both Hormone-B and Hormone-C versions of the Ka-25.

MBB BO 105L

Country of Origin: Federal Germany.
Type: Five/six-seat light utility helicopter.
Power Plant: Two 550 shp Allison 250-C28C turboshafts.
Performance: Max. speed, 168 mph (270 km/h) at sea level; max. cruise, 157 mph (252 km/h) at sea level; max. climb, 1,970 ft/min (10 m/sec); hovering ceiling (in ground effect), 13,120 ft (4 000 m), (out of ground effect), 11,280 ft (3 440 m); range, 286 mls (460 km).
Weights: Empty, 2,756 lb (1 250 kg); max. take-off, 5,291 lb (2 400 kg), (with external load), 5,512 lb (2 500 kg).
Dimensions: Rotor diam, 32 ft 3½ in (9,84 m); fuselage length, 28 ft 1 in (8,56 m).
Notes: The BO 105L is a derivative of the BO 105CB (see 1979 edition) with uprated transmission and more powerful turboshaft for "hot-and-high" conditions. It is otherwise similar to the BO 105CBS Twin Jet II (420 shp Allison 250-C20B) which was continuing in production at the beginning of 1983, when more than 700 BO 105s (all versions) had been delivered and production was running at 10–12 monthly, and licence assembly was being undertaken in Indonesia, the Philippines and Spain. Deliveries to the Federal German Army of 227 BO 105M helicopters for liaison and observation tasks commenced late 1979, and deliveries of 212 HOT-equipped BO 105s (illustrated) for the anti-armour role began on 4 December 1980. The latter have uprated engines and transmission systems.

MBB-KAWASAKI BK 117

Countries of Origin: Federal Germany and Japan.
Type: Multi-purpose eight-to-twelve-seat helicopter.
Power Plant: Two 600 shp Avco Lycoming LTS 101-650B-1 turboshafts.
Performance: Max. speed, 171 mph (275 km/h) at sea level; cruise, 164 mph (264 km/h) at sea level; max. climb, 1,970 ft/min (10 m/sec); hovering ceiling (in ground effect), 13,450 ft (4 100 m), (out of ground effect), 10,340 ft (3 150 m); range (max. payload), 339 mls (545,4 km).
Weights: Empty, 3,351 lb (1 520 kg); max. take-off, 6,173 lb (2 800 kg).
Dimensions: Rotor diam, 36 ft 1 in (11,00 m); fuselage length, 32 ft 5 in (9,88 m).
Notes: The BK 117 is a co-operative development between Messerschmitt-Bölkow-Blohm and Kawasaki, the first of two flying prototypes commencing its flight test programme on 13 June 1979 (in Germany), with the second following on 10 August (in Japan). A decision to proceed with series production was taken in 1980, with first flying on 24 December 1981, and production deliveries commencing first quarter of 1983. Some 130 BK 117s had been ordered by the beginning of 1983. MBB is responsible for the main and tail rotor systems, tail unit and hydraulic components, while Kawasaki is responsible for production of the fuselage, undercarriage, transmission and some other components. Several military versions are currently proposed.

MIL MI-8 (HIP)

Country of Origin: USSR.
Type: Assault transport helicopter.
Power Plant: Two 1,700 shp Isotov TV2-117A turboshafts.
Performance: Max. speed, 161 mph (260 km/h) at 3,280 ft (1 000 m), 155 mph (250 km/h) at sea level; max. cruise, 140 mph (225 km/h); hovering ceiling (in ground effect), 6,233 ft (1 900 m), (out of ground effect), 2,625 ft (800 m); range (standard fuel), 290 mls (465 km).
Weights: (Hip-C) Empty, 14,603 lb (6 624 kg); normal loaded, 24,470 lb (11 100 kg); max. take-off, 26,455 lb (12 000 kg).
Dimensions: Rotor diam, 69 ft 10$\frac{1}{4}$ in (21,29 m); fuselage length, 60 ft 0$\frac{3}{4}$ in (18,31 m).
Notes: Currently being manufactured at a rate of 700–800 annually, with 6,000–7,000 delivered for civil and military use since its debut in 1961, the Mi-8 is numerically the most important Soviet helicopter. Current military versions include the Hip-C basic assault transport, the Hip-D with additional antennae and podded equipment for electronic tasks, the Hip-E and the Hip-F, the former carrying up to six rocket pods and four Swatter IR-homing anti-armour missiles, and the latter carrying six Sagger wire-guided anti-armour missiles. The Mi-8 can accommodate 24 troops or 12 stretchers, and most have a 12,7-mm machine gun in the nose. Commercial models include the basic 28–32-passenger model and the Mi-8T utility version. An enhanced version, the Mi-17, is described on page 241.

MIL MI-14 (HAZE-A)

Country of Origin: USSR.
Type: Amphibious anti-submarine helicopter.
Power Plant: Two 1,500 shp Isotov TV-2 turboshafts.
Performance: (Estimated) Max. speed, 143 mph (230 km/h); max. cruise, 130 mph (210 km/h); hovering ceiling (in ground effect), 5,250 ft (1 600 m), (out of ground effect), 2,295 ft (700 m); tactical radius, 124 mls (200 km).
Weights: (Estimated) Max. take-off, 26,455 lb (12 000 kg).
Dimensions: Rotor diam, 69 ft $10\frac{1}{4}$ in (21,29 m); fuselage length, 59 ft 7 in (18,15 m).
Notes: The Mi-14 amphibious anti-submarine warfare helicopter, which serves with shore-based elements of the Soviet Naval Air Force, is a derivative of the Mi-8 (see page 239) with essentially similar power plant and dynamic components, and much of the structure is common between the two helicopters. New features include the boat-type hull, outriggers which, housing the retractable lateral twin-wheel undercarriage members, incorporate water rudders, a search radar installation beneath the nose and a sonar "bird" beneath the tailboom root. The Mi-14 may presumably be used for over-the-horizon missile targeting and for such tasks as search and rescue. It may also be assumed that the Mi-14 possesses a weapons bay for ASW torpedoes, nuclear depth charges and other stores. This amphibious helicopter reportedly entered service in 1975 and about 120 were in Soviet Navy service by the beginning of 1983.

MIL MI-17

Country of Origin: USSR.
Type: Medium transport helicopter.
Power Plant: Two 1,900 shp Isotov TV3-117MT turboshafts.
Performance: (At 28,660 lb/13 000 kg) Max. speed, 162 mph (260 km/h); max. continuous cruise, 149 mph (240 km/h) at sea level; hovering ceiling (at 24,250 lb/ 11 000 kg out of ground effect), 5,800 ft (1 770 m); max. range, 590 mls (950 km).
Weights: Empty, 15,652 lb (7 100 kg); normal loaded, 24,250 lb (11 000 kg); max. take-off, 28,660 lb (13 000 kg).
Dimensions: Rotor diam, 69 ft $10\frac{1}{4}$ in (21,29 m); fuselage length, 60 ft $5\frac{1}{4}$ in (18,42 m).
Notes: The Mi-17 medium-lift helicopter is essentially a more powerful and modernised derivative of the late fifties technology Mi-8 (see page 239). The airframe and rotor are essentially unchanged, apart from some structural reinforcement of the former, but higher-performance turboshafts afford double the normal climb rate and out-of-ground-effect hover ceiling of the earlier helicopter, and increase permissible maximum take-off weight. The Mi-17 has a crew of two–three and can accommodate 24 passengers, 12 casualty stretchers or up to 8,818 lb (4 000 kg) of freight. The TV3-117 turboshafts utilised by the Mi-17 are also believed to be installed in late production version of the military Mi-8 (eg, the Hip-E and Hip-F). Externally, the Mi-17 is virtually indistinguishable from its precursor, the Mi-8, apart from marginally shorter engine nacelles.

MIL-MI-24 (HIND-D)

Country of Origin: USSR.
Type: Assault and anti-armour helicopter.
Power Plant: Two 2,200 shp Isotov TV3-117 turboshafts.
Performance: (Estimated) Max. speed, 170–180 mph (273–290 km/h) at 3,280 ft (1 000 m); max. cruise, 145 mph (233 km/h); max. inclined climb rate, 3,000 ft/min (15,24 m/sec).
Weights: (Estimated) Normal take-off, 22,000 lb (10 000 kg).
Dimensions: (Estimated) Rotor diam, 55 ft 0 in (16,76 m); fuselage length, 55 ft 6 in (16,90 m).
Notes: By comparison with the Hind-A version of the Mi-24 (see 1977 edition), the Hind-D embodies a redesigned forward fuselage and is optimised for the gunship role, having tandem stations for the weapons operator (in nose) and pilot. The Hind-D can accommodate eight fully-equipped troops, has a barbette-mounted four-barrel rotary-type 12,7-mm cannon beneath the nose and can carry up to 2,800 lb (1 275 kg) of ordnance externally, including four AT-2 Swatter IR-homing anti-armour missiles and four pods each with 32 57-mm rockets. It has been exported to Afghanistan, Algeria, Bulgaria, Czechoslovakia, East Germany, Hungary, Iraq, Libya, Poland and South Yemen. The Hind-E is similar but has provision for four laser-homing tube-launched Spiral anti-armour missiles, may be fitted with a twin-barrel 23-mm cannon on each side of the fuselage and embodies some structural hardening, steel and titanium being substituted for aluminium in certain critical components.

MIL MI-26 (HALO)

Country of Origin: USSR.
Type: Military and commercial heavy-lift helicopter.
Power Plant: Two 11,400 shp Lotarev D-136 turboshafts.
Performance: Max. speed, 183 mph (295 km/h); normal cruise, 158 mph (255 km/h); hovering ceiling (in ground effect), 14,765 ft (4 500 m), (out of ground effect), 5,905 ft (1 800 m); range (at 109,127 lb/49 500 kg), 310 mls (500 km), (at 123,457 lb/56 000 kg), 497 mls (800 km).
Weights: Empty, 62,169 lb (28 200 kg); normal load, 109,227 lb (49 500 kg); max. take-off, 123,457 lb (56 000 kg).
Dimensions: Rotor diam, 104 ft $11\frac{7}{8}$ in (32,00 m); fuselage length (nose to tail rotor), 110 ft $7\frac{3}{4}$ in (33,73 m).
Notes: The heaviest and most powerful helicopter ever flown, the Mi-26 first flew as a prototype on 14 December 1977, production of pre-series machines commencing in 1980, and preparations for full-scale production having begun in 1981. Featuring an innovative eight-bladed main rotor and carrying a flight crew of five, the Mi-26 has a max. internal payload of 44,090 lb (20 000 kg). The freight hold is larger than that of the fixed-wing Antonov An-12 transport and at least 70 combat-equipped troops or 40 casualty stretchers can be accommodated. Although allegedly developed to a civil requirement, the primary role of the Mi-26 is obviously military and it is anticipated that the Soviet Air Force will achieve initial operational capability with the series version in 1984–85. During the course of 1982, the Mi-26 established a number of new international payload-to-height records.

SIKORSKY CH-53E SUPER STALLION

Country of Origin: USA.
Type: Amphibious assault transport helicopter.
Power Plant: Three 4,380 shp General Electric T64-GE-415 turboshafts.
Performance: (At 56,000 lb/25 400 kg) Max. speed, 196 mph (315 km/h) at sea level; cruise, 173 mph (278 km/h) at sea level; max. inclined climb, 2,750 ft/min (13,97 m/sec); hovering ceiling (in ground effect), 11,550 ft (3 520 m), (out of ground effect), 9,500 ft (2 895 m); range, 1,290 mls (2 075 km).
Weights: Empty, 33,226 lb (15 071 kg); max. take-off, 73,500 lb (33 339 kg).
Dimensions: Rotor diam, 79 ft 0 in (24,08 m); fuselage length, 73 ft 5 in (22,38 m).
Notes: The CH-53E is a growth version of the CH-53D Sea Stallion (see 1974 edition) embodying a third engine, an uprated transmission system, a seventh main rotor blade and increased rotor diameter. The first of two prototypes was flown on 1 March 1974, and the first of two pre-production examples followed on 8 December 1975, successive production orders totalling 75 helicopters to be divided between the US Navy and US Marine Corps by beginning of 1983, the first production Super Stallion having made its first flight on 13 December 1980. The CH-53E can accommodate up to 55 troops in a high-density seating arrangement. Fleet deliveries began mid-1981, and the 35th Super Stallion is being completed as the MH-53E mine countermeasures production prototype.

SIKORSKY S-70 (UH-60A) BLACK HAWK

Country of Origin: USA.
Type: Tactical transport helicopter.
Power Plant: Two 1,543 shp General Electric T700-GE-700 turboshafts.
Performance: Max. speed, 224 mph (360 km/h) at sea level; cruise, 166 mph (267 km/h); vertical climb rate, 450 ft/min (2,28 m/sec); hovering ceiling (in ground effect), 10,000 ft (3 048 m), (out of ground effect), 5,800 ft (1 758 m); endurance, 2·3-3·0 hrs.
Weights: Design gross, 16,500 lb (7 485 kg); max. take-off, 22,000 lb (9 979 kg).
Dimensions: Rotor diam, 53 ft 8 in (16,23 m); fuselage length, 50 ft 0¾ in (15,26 m).
Notes: The Black Hawk was winner of the US Army's UTTAS (Utility Tactical Transport Aircraft System) contest, and contracts had been announced by beginning of 1982 for 342 examples. The first of three YUH-60As was flown on 17 October 1974, and a company-funded fourth prototype flew on 23 May 1975. The Black Hawk is primarily a combat assault squad carrier, accommodating 11 fully-equipped troops. Three variants under development at the beginning of 1983 were the EC-60A ECM model, the EC-60B for target acquisition and the HH-60D Night Hawk rescue helicopter. The USAF has a requirement for 243 HH-60Ds. The first production deliveries to the US Army were made in June 1979, with 350 delivered by beginning of 1983 against requirement for 1,107.

SIKORSKY S-70L (SH-60B) SEA HAWK

Country of Origin: USA.
Type: Shipboard multi-role helicopter.
Power Plant: Two 1,690 shp General Electric T700-GE-401 turboshafts.
Performance: (At 20,244 lb/9 183 kg) Max. speed, 167 mph (269 km/h) at sea level; max. cruising speed, 155 mph (249 km/h) at 5,000 ft (1 525 m); max. vertical climb, 1,192 ft/min (6,05 m/sec); time on station (at radius of 57 mls/92 km), 3 hrs 52 min.
Weights: Empty equipped, 13,678 lb (6 204 kg); max. take-off, 21,844 lb (9 908 kg).
Dimensions: Rotor diam, 53 ft 8 in (16,36 m); fuselage length, 50 ft 0¾ in (15,26 m).
Notes: Winner of the US Navy's LAMPS (Light Airborne Multi-Purpose System) Mk III helicopter contest, the SH-60B is intended to fulfil both anti-submarine warfare (ASW) and anti-ship surveillance and targeting (ASST) missions and the first of five prototypes was flown on 12 December 1979, and the last on 14 July 1980. Evolved from the UH-60A (see page 245), the SH-60B is intended to serve aboard DD-963 destroyers, DDG-47 Aegis cruisers and FFG-7 guided-missile frigates as an integral extension of the sensor and weapon system of the launching vessel. The US Navy has a requirement for 204 LAMPS III category helicopters, with delivery of first seven from April 1983, and 195 simplified SH-60Cs from 1987.

SIKORSKY S-76 MK II

Country of Origin: USA.
Type: Fourteen-seat commercial transport helicopter.
Power Plant: Two 700 shp Allison 250-C30 turboshafts.
Performance: Max. speed, 179 mph (288 km/h); max. cruise, 167 mph (268 km/h); range cruise, 145 mph (233 km/h); hovering ceiling (in ground effect), 5,100 ft (1 524 m), (out of ground effect), 1,400 ft (427 m); range (full payload and 30 min reserve), 460 mls (740 km).
Weights: Empty 5,600 lb (2 540 kg); max. take-off, 10,300 lb (4 672 kg).
Dimensions: Rotor diam, 44 ft 0 in (13,41 m); fuselage length, 43 ft 4½ in (13,22 m).
Notes: The first of four prototypes of the S-76 flew on 13 March 1977, and customer deliveries commenced 1979, with 200 being delivered by the beginning of 1983, when a production rate of two per month was being maintained. The S-76 is unique among Sikorsky commercial helicopters in that conceptually it owes nothing to an existing military model, although it has been designed to conform with appropriate military specifications and military customers were included among contracts for helicopters of this type that had been ordered by the beginning of 1983. The S-76 may be fitted with extended-range tanks, cargo hook and rescue hoist. The main rotor is a scaled-down version of that used by the UH-60. All S-76s were being modified to Mk II standards from mid-1982.

SIKORSKY S-76 UTILITY

Country of Origin: USA.
Type: General-purpose military utility helicopter.
Power Plant: Two 700 shp Allison 250-C30S or 1,000 shp Pratt & Whitney PT6B-36 turboshafts.
Performance: (Allison 250-C30S) Max. speed, 178 mph (287 km/h); average cruise, 173 mph (278 km/h); range (with 30 min reserves), 465 mls (749 km).
Weights: Empty, 5,610 lb (2 545 kg); max. take-off, 10,300 lb (4 672 kg).
Dimensions: Rotor diam, 44 ft 0 in (13,41 m); fuselage length, 43 ft 5 in (13,21 m).
Notes: A derivative of the S-76 Mk II (see page 247), the S-76 Utility, which entered flight test early 1982 and completed weapons qualifications in August of that year, is a multi-role helicopter suitable for use as a gunship, utility transport accommodating up to 12 troops, or for scouting and reconnaissance missions. It has a high-strength fuel tank, armoured pilots' seats and sliding doors in both sides of the fuselage, a fixed undercarriage being optional. In armed configuration, it has weapons pylons and provision for door-mounted weapons. Podded 7,62-mm or 0·5-in (12,7-mm) machine guns, or 7,62-mm miniguns may be fitted to the outrigger-type pylons, optional ordnance including pods of 19 2·75-in (7,0-mm) folding-fin rockets or up to eight TOW anti-armour missiles. A version is on offer in emergency medical configuration, and several dedicated variants are proposed.

WESTLAND SEA KING

Country of Origin: United Kingdom (US licence).
Type: Anti-submarine warfare and search-and-rescue helicopter.
Power Plant: Two 1,660 shp Rolls-Royce Gnome H.1400-1 turboshafts.
Performance: Max. speed, 143 mph (230 km/h); max. continuous cruise at sea level, 131 mph (211 km/h); hovering ceiling (in ground effect), 5,000 ft (1 525 m), (out of ground effect), 3,200 ft (975 m); range (standard fuel), 764 mls (1 230 km), (auxiliary fuel), 937 mls (1 507 km).
Weights: Empty equipped (ASW), 13,672 lb (6 201 kg), (SAR), 12,376 lb (5 613 kg); max. take-off, 21,000 lb (9 525 kg).
Dimensions: Rotor diam, 62 ft 0 in (18,90 m); fuselage length, 55 ft 9¾ in (17,01 m).
Notes: The Sea King Mk 2 is an uprated version of the basic ASW and SAR derivative of the licence-built S-61D (see 1982 edition), the first Mk 2 being flown on 30 June 1974, and being one of 10 Sea King Mk 50s ordered by the Australian Navy. Twenty-one delivered to the Royal Navy as Sea King HAS Mk 2s, and 15 examples of a SAR version to the RAF as Sea King HAR Mk 3s. Current production version is the Sea King HAS Mk 5, delivery of 17 to Royal Navy having commenced October 1980. All HAS Mk 2s being brought up to Mk 5 standards and five have been fitted with Thorn-EMI searchwater radar (as illustrated) for airborne early warning duty.

WESTLAND COMMANDO

Country of Origin: United Kingdom (US licence).
Type: Tactical transport helicopter.
Power Plant: Two 1,590 shp Rolls-Royce Gnome H-1400-1 turboshafts.
Performance: (Mk 2 at 21,000 lb/9 526 kg) Max. speed, 140 mph (225 km/h) at sea level; normal operating speed, 129 mph (207 km/h) at sea level; initial inclined climb, 2,020 ft/min (10,3 m/sec); hovering ceiling (in ground effect), 5,000 ft (1 525 m), (out of ground effect), 3,200 ft (975 m); range (with max. payload), 276 mls (445 km).
Weights: Basic equipped, 11,174 lb (5 069 kg); max. take-off, 21,000 lb (9 526 kg).
Dimensions: Rotor diam, 62 ft 0 in (18,90 m); fuselage length, 55 ft 9¾ in (17,01 m).
Notes: The Commando is a Westland-developed land-based army support helicopter derivative of the Sea King (see page 249), the interim Mk 1 having minimum changes from the ASW helicopter. The first of five Mk 1s was flown on 12 September 1973, subsequent examples being built to Mk 2 standard with the uprated Gnome turboshafts selected for the Sea King Mk 50. The five Mk 1s and 23 Mk 2s were supplied to Egypt and four Mk 2s to Qatar, and 15 similar helicopters were supplied (as Sea King HC Mk 4s—illustrated) to the Royal Navy commando squadrons. In mid-1982, it was announced that a further eight were to be ordered. The Mk 3 (with sponsons and retractable undercarriage) flew on 14 June 1982, eight having been ordered by Qatar.

WESTLAND WG 13 LYNX

Country of Origin: United Kingdom.
Type: Multi-purpose, ASW and transport helicopter.
Power Plant: Two 900 shp Rolls-Royce BS. 360-07-26 Gem 100 turboshafts.
Performance: Max. speed, 207 mph (333 km/h); max. continuous sea level cruise, 170 mph (273 km/h); max. inclined climb, 1,174 ft/min (11,05 m/sec); hovering ceiling (out of ground effect), 12,000 ft (3 660 m); max. range (internal fuel), 391 mls (629 km); max. ferry range (auxiliary fuel), 787 mls (1 266 km).
Weights: (HAS Mk 2) Operational empty, 6,767–6,999 lb (3 069–3 179 kg); max. take-off, 9,500 lb (4 309 kg).
Dimensions: Rotor diam, 42 ft 0 in (12,80 m); fuselage length, 39 ft 1¼ in (11,92 m).
Notes: The first of 13 development Lynxes was flown on 21 March 1971, with the first production example (an HAS Mk 2) flying on 10 February 1976. By the beginning of 1983 production rate was nine per month and a total of 310 was on order, including 40 for the French Navy, 80 for the Royal Navy, 114 for the British Army, 10 for the Argentine Navy, eight for the Danish Navy, 12 for the German Navy, nine for the Brazilian Navy, six for Norway, 24 for the Netherlands Navy, six for the Nigerian Navy and three of a general-purpose version for Qatar. The Lynx AH MK 1 is the British Army's general utility version and the Lynx HAS Mk 2 is the ASW version for the Royal Navy. Eighteen of the Dutch and 14 of the French Lynx have uprated engines.

WESTLAND WG 30-100

Country of Origin: United Kingdom.
Type: Transport and utility helicopter.
Power Plant: Two 1,265 shp Rolls-Royce Gem 60-1 turboshafts.
Performance: Max. speed (at 10,500 lb/4 763 kg), 163 mph (263 km/h) at 3,000 ft (915 m); hovering ceiling (in ground effect), 7,200 ft (2 195 m), (out of ground effect), 5,000 ft (1 525 m); range (seven passengers), 426 mls (686 km).
Weights: Operational empty (typical), 6,880 lb (3 120 kg); max. take-off, 12,800 lb (5 806 kg).
Dimensions: Rotor diam, 43 ft 8 in (13,31 m); fuselage length, 47 ft 0 in (14,33 m).
Notes: The WG 30, flown for the first time on 10 April 1979, is a private venture development of the Lynx (see page 251) featuring an entirely new fuselage offering a substantial increase in capacity. Aimed primarily at the multi-role military helicopter field, the WG 30 has a crew of two and in the transport role can carry 17–22 passengers. Commitment to the WG 30 at the time of closing for press covers initial production batch of 20, deliveries of which began January 1982. British Airways has ordered two and a contract for six has been placed by the US-based Airspur Airline which also has options on a further 15. The WG 30 utilises more than 85% of the proven systems of the WG 13 Lynx, and the WG 30-200 differs from the -100 in having General Electric CT7-2 turboshafts.

ACKNOWLEDGEMENTS

The author wishes to record his thanks to the following sources of copyright photographs appearing in this volume: Avio data, page 6; Jan Čech, page 112; Aviation Magazine International, page 214; Israir, pages 92, 96, 130 and 217, Stephen Peltz, pages, 56, 110, 160, 188 and 192, and Brian M. Service, 241.

INDEX OF AIRCRAFT TYPES